ENVIRONMENTAL REMEDIATION TECHNOLOGIES, REGULATIONS AND SAFETY

THE GREAT LAKES RESTORATION INITIATIVE

ENVIRONMENTAL REMEDIATION TECHNOLOGIES, REGULATIONS AND SAFETY

Additional books in this series can be found on Nova's website under the Series tab.

Additional e-books in this series can be found on Nova's website under the e-book tab.

ENVIRONMENTAL REMEDIATION TECHNOLOGIES, REGULATIONS AND SAFETY

THE GREAT LAKES RESTORATION INITIATIVE

ÓSKAR STROBEL
EDITOR

nova publishers
New York

For permission to use material from this book please contact us:
Telephone 631-231-7269; Fax 631-231-8175
Web Site: http://www.novapublishers.com

Library of Congress Cataloging-in-Publication Data

ISBN: 978-1-63117-525-1

Published by Nova Science Publishers, Inc. † New York

CONTENTS

Preface		**vii**
Chapter 1	The Great Lakes Restoration Initiative: Background and Issues *Pervaze A. Sheikh*	**1**
Chapter 2	Great Lakes Restoration Initiative: Further Actions Would Result in More Useful Assessments and Help Address Factors That Limit Progress *United States Government Accountability Office*	**29**
Chapter 3	Great Lakes Restoration Initiative Action Plan Fiscal Years 2010-2014 *Great Lakes Interagency Task Force*	**119**
Index		**177**

PREFACE

This books provides the reader with a background on and issues of the Great Lakes restoration initiative, as well as the action plan for the years 2010-2014.

Chapter 1 – The Great Lakes ecosystem is recognized by many as an international natural resource that has been altered by human activities and climate variability. These alterations have led to degraded water quality, diminished habitat, lower native fish and wildlife populations, and an altered ecosystem. In response, the federal governments of the United States and Canada and the state and provincial governments in the Great Lakes basin are implementing several restoration activities. These activities range from mitigating the harmful effects of toxic substances in lake waters to restoring fish habitat.

Most laws and efforts in the past addressed specific issues in the Great Lakes; a few addressed issues at the ecosystem level. This caused the Government Accountability Office and others to express the need for initiating and implementing a comprehensive approach for restoring the Great Lakes ecosystem. In 2010, the Great Lakes Restoration Initiative (GLRI) was proposed and implemented by the Obama Administration. The aim of GLRI is to restore the Great Lakes ecosystem under one initiative. Specifically, the GLRI is to restore and maintain the chemical, physical and biological integrity of the Great Lakes Basin Ecosystem by directing activities to address five focus areas: (1) toxic substances and Areas of Concern (these are areas in the Great Lakes that are environmentally degraded); (2) invasive species; (3) nearshore health and nonpoint source pollution; (4) habitat and wildlife protection and restoration; and (5) accountability, monitoring, evaluation, communication, and partnerships.

The Environmental Protection Agency (EPA) is the lead federal agency for implementing and administering GLRI. The EPA has received authority to distribute appropriated funds to several federal agencies, which then undertake restoration activities and projects. The EPA also administers grant programs to fund nonfederal projects and activities related to restoration. An interagency Great Lakes Task Force oversees the implementation of GLRI and created a strategy to guide restoration. The strategy (referred to as the Action Plan) provides a framework for restoring the Great Lakes ecosystem under GLRI from 2010 through 2014. For each focus area under the GLRI, the Action Plan provides a problem statement, a set of goals, interim objectives, progress measures, final targets, and principal activities for restoring the ecosystem. Restoration activities are being done under existing federal authorities. The GLRI has received approximately $1.37 billion in appropriated funds since FY2010.

The scope and scale of this restoration initiative have led some to question its direction and duration. The GLRI does not specify what a restored ecosystem might look like, nor does it estimate how long restoration activities will need to be conducted, and how much restoration might cost. Some other questions surrounding this initiative include how the GLRI is governed and how federal and state restoration efforts are coordinated. Furthermore, GLRI remains an administrative initiative; there is no law that specifically authorizes GLRI, though Congress has appropriated funds to implement the program. Congress might consider these questions in oversight hearings or in legislation during the 113th Congress. Companion bills have been introduced in the 113th Congress to address GLRI. S. 1232 and H.R. 2773 would establish an administrative and management structure for restoration activities in the Great Lakes, authorize GLRI and appropriations for its implementation, specify the scope and function of GLRI, and authorize the coordinating role of the Great Lakes Interagency Task Force.

Chapter 2 – The Great Lakes contain about 84 percent of North America's surface freshwater and provide economic and recreational benefits in the Great Lakes Basin. However, the Great Lakes face significant stresses—such as toxic pollution—that have caused ecological and economic damage to the region.

Approximately $1.3 billion has been appropriated to the GLRI, created in fiscal year 2010, which an interagency Task Force of 11 federal agencies, chaired by the EPA Administrator, oversees. In 2010, the Task Force issued an Action Plan for fiscal years 2010 to 2014 to develop a comprehensive approach to restoring the health of the Great Lakes ecosystem. GAO was asked to review the GLRI. This report examines (1) how the GLRI is implemented by the Task Force agencies and other stakeholders, (2) the methods that EPA has in place to assess GLRI progress, (3) the progress identified by the Task Force agencies and nonfederal stakeholders, and (4) the views of nonfederal stakeholders on factors, if any, that may affect or limit GLRI progress. GAO analyzed the Action Plan, surveyed 205 non-federal recipients of GLRI funding, and interviewed Task Force agency officials and nonfederal stakeholders.

Chapter 3 – The lands and waters of the Great Lakes are like no other place. In a world where fresh surface water is increasingly in demand, the region contains some 20 percent of it. At a time when people are not looking as much to faraway places for respite, the Great Lakes offer some of the most majestic natural shorescapes on the planet to accommodate them. As a result, people are reconnecting to their beaches, wide-open waters, petro glyphed bluffs, dune ranges and tumbling tributaries like never before. Likewise, these same resources have served as the raw material to build some of the Earth's most legendary cities, create jobs to support families, and contribute to the largest economy in the history of the world.

Still, our expectation that the Great Lakes will continue to meet these needs has resulted in lost flora, fauna, soil, and air and water quality to the point where the ecosystem is showing signs of severe stress and its ability to keep up with these demands is in doubt.

While in the past we have worked to minimize harm, public demand for a new standard of care is surging. That standard of care is that we must leave the Great Lakes better for the next generation than the condition in which we inherited them. We must continue to go beyond minimizing harm to proactively rehabilitating the Great Lakes. Only then will they be able to keep providing jobs, recreation and sanctuary.

In: The Great Lakes Restoration Initiative
Editor: Óskar Strobel
ISBN: 978-1-63117-525-1

Chapter 1

THE GREAT LAKES RESTORATION INITIATIVE: BACKGROUND AND ISSUES*

Pervaze A. Sheikh

SUMMARY

The Great Lakes ecosystem is recognized by many as an international natural resource that has been altered by human activities and climate variability. These alterations have led to degraded water quality, diminished habitat, lower native fish and wildlife populations, and an altered ecosystem. In response, the federal governments of the United States and Canada and the state and provincial governments in the Great Lakes basin are implementing several restoration activities. These activities range from mitigating the harmful effects of toxic substances in lake waters to restoring fish habitat.

Most laws and efforts in the past addressed specific issues in the Great Lakes; a few addressed issues at the ecosystem level. This caused the Government Accountability Office and others to express the need for initiating and implementing a comprehensive approach for restoring the Great Lakes ecosystem. In 2010, the Great Lakes Restoration Initiative (GLRI) was proposed and implemented by the Obama Administration. The aim of GLRI is to restore the Great Lakes ecosystem under one initiative. Specifically, the GLRI is to restore and maintain the chemical,

* This is an edited, reformatted and augmented version of a Congressional Research Service publication, CRS Report for Congress R43249, from www.crs.gov, prepared for Members and Committees of Congress, dated September 30, 2013.

physical and biological integrity of the Great Lakes Basin Ecosystem by directing activities to address five focus areas: (1) toxic substances and Areas of Concern (these are areas in the Great Lakes that are environmentally degraded); (2) invasive species; (3) nearshore health and nonpoint source pollution; (4) habitat and wildlife protection and restoration; and (5) accountability, monitoring, evaluation, communication, and partnerships.

The Environmental Protection Agency (EPA) is the lead federal agency for implementing and administering GLRI. The EPA has received authority to distribute appropriated funds to several federal agencies, which then undertake restoration activities and projects. The EPA also administers grant programs to fund nonfederal projects and activities related to restoration. An interagency Great Lakes Task Force oversees the implementation of GLRI and created a strategy to guide restoration. The strategy (referred to as the Action Plan) provides a framework for restoring the Great Lakes ecosystem under GLRI from 2010 through 2014. For each focus area under the GLRI, the Action Plan provides a problem statement, a set of goals, interim objectives, progress measures, final targets, and principal activities for restoring the ecosystem. Restoration activities are being done under existing federal authorities. The GLRI has received approximately $1.37 billion in appropriated funds since FY2010.

The scope and scale of this restoration initiative have led some to question its direction and duration. The GLRI does not specify what a restored ecosystem might look like, nor does it estimate how long restoration activities will need to be conducted, and how much restoration might cost. Some other questions surrounding this initiative include how the GLRI is governed and how federal and state restoration efforts are coordinated. Furthermore, GLRI remains an administrative initiative; there is no law that specifically authorizes GLRI, though Congress has appropriated funds to implement the program. Congress might consider these questions in oversight hearings or in legislation during the 113th Congress. Companion bills have been introduced in the 113th Congress to address GLRI. S. 1232 and H.R. 2773 would establish an administrative and management structure for restoration activities in the Great Lakes, authorize GLRI and appropriations for its implementation, specify the scope and function of GLRI, and authorize the coordinating role of the Great Lakes Interagency Task Force.

INTRODUCTION

The Great Lakes ecosystem is the largest system of fresh surface water in the world. The watershed covers approximately 300,000 square miles and is shared by eight U.S. states (Illinois, Indiana, Michigan, Minnesota, New York, Ohio, Pennsylvania, and Wisconsin) and one Canadian province (Ontario). (See *Figure 1*.) The Great Lakes contain nearly 90% of the surface freshwater of the United States and 20% of the surface freshwater of the world. In the last several decades, agricultural activity throughout the basin, and urban and industrial development concentrated along the shoreline, have degraded water quality in the Great Lakes, posing potential threats to the ecosystem.

Source: U.S. Army Corps of Engineers, Detroit District.

Figure 1. The Great Lakes Basin.

Development has also led to changes in terrestrial and aquatic habitats, the introduction of nonnative species, the contamination of sediments, and the listing of more than 50 threatened or endangered species in the ecosystem.[1] In response to deteriorating conditions in the ecosystem, the federal governments of the United States and Canada and the state and provincial governments in the Great Lakes basin have implemented several restoration activities.[2]

An estimated 40 million people rely on the Great Lakes ecosystem for jobs, drinking water, and recreation, among other things. In economic terms, the present-value benefit in 2007 of restoring the Great Lakes was estimated to be $50 billion (in direct and indirect benefits) over the long term.[3] The ultimate cost for restoring the Great Lakes is unclear, but the value of the potential benefits of restoration has caused, in part, several in Congress to regard restoring the Great Lakes a priority. The U.S. Congress has played a role in restoration efforts in the Great Lakes and enacted more than 30 laws over several years focused on restoring aspects of the Great Lakes ecosystem. These laws have authorized activities ranging from mitigating the harmful effects of toxic substances on water quality to mitigating damages caused by invasive species. Most laws address specific issues in the Great Lakes; yet few address the entire ecosystem. Over the years, several stakeholders have expressed the need for initiating and implementing a comprehensive approach for restoring the Great Lakes ecosystem.[4]

Restoration efforts in the Great Lakes have been implemented over several decades by the federal government, states, and local stakeholders. Efforts have focused on specific aspects of the ecosystem, but rarely the entire ecosystem. For example, the Great Lakes Water Quality Agreement (GLWQA), initially signed in 1972 and then revised several times afterwards, addresses water quality characteristics in the Great Lakes, but not habitat issues.[5] After several years of restoration, the Government Accountability Office (GAO) concluded that progress in restoring the Great Lakes was slow and restoration efforts were too loosely organized. [6] Specific concerns included the slow rate of cleaning up toxic sediments, insufficient governance to provide direction for ecosystem restoration activities, and lack of a comprehensive plan to guide restoration of the Great Lakes ecosystem.[7]

In 2004 a federal Great Lakes Interagency Task Force (hereinafter referred to as the Task Force)[8] was created to provide strategic direction for Great Lakes policies on restoration and to form a regional collaboration of stakeholders interested in restoring the Great Lakes ecosystem. This collaboration was termed the Great Lakes Regional Collaboration. It consisted of over 1,500 stakeholders and released the Great Lakes Regional Collaboration Strategy (hereinafter referred to as the Strategy).[9] The Strategy recommended implementing a series of actions and activities to start the restoration of the Great Lakes ecosystem over a five-year period from 2006 to 2011. The Strategy encompassed eight issue areas: aquatic invasive species, fish and wildlife habitat (habitat/species), coastal health, contaminated sediments, nonpoint source pollution, toxic pollutants, indicators and information, and sustainable development. The total cost of implementing the Strategy was estimated to be $20 billion over the five-year period. The Strategy (as a whole) is not being implemented, although several restoration programs included in the Strategy are underway or ongoing in the Great Lakes Restoration Initiative (GLRI) and other federal and state activities.

Great Lakes Restoration Initiative

The GLRI was proposed in 2009 by the Obama Administration, and implemented in 2010. The GLRI is derived from the Strategy and aims to be consistent with the Strategy and GLWQA. The goal of the GLRI is to restore and maintain the chemical, physical and biological integrity of the Great Lakes Basin Ecosystem by directing activities to address five focus areas:[10]

- toxic substances and Areas of Concern;[11]
- invasive species;
- nearshore health and nonpoint source pollution;
- habitat and wildlife protection and restoration;
- accountability, monitoring, evaluation, communication, and partnerships.

Federal efforts to restore the Great Lakes ecosystem are coordinated by the Task Force. The EPA, serving as chair of the Task Force, is the lead federal agency for implementing and administering the GLRI. In appropriations laws from FY2010 to FY2013, the EPA has been given authority to receive and distribute congressionally appropriated funds to several federal agencies, which then undertake restoration activities and projects in the Great Lakes. EPA also implements restoration activities that are funded by the GLRI through the Great Lakes National Program Office (GLNPO). Restoration activities are being done under existing federal authorities that address restoration in the Great Lakes. There is no single law that specifically authorizes the GLRI as a restoration initiative for the Great Lakes.

The implementation of the GLRI is being guided by recommendations from a Great Lakes Advisory Board (GLAB). The GLAB provides advice on Great Lakes protection and restoration policy, long term goals of protection and restoration, annual priorities to protect and restore the Great Lakes, and issues addressed by the Task Force.[12] GLAB consists of 18 members who represent nonfederal stakeholders (e.g., non-governmental organizations, state agencies, tribal interests, and universities, among others) in the GLRI. GLAB is also expected to provide advice on the implementation of the Great Lakes Water Quality Agreement.[13]

GLRI Action Plan

The implementation of GLRI is also being guided by a Great Lakes Restoration Initiative Action Plan (hereinafter referred to as the Action Plan), which was created by the Task Force.[14] The Action Plan provides a framework for restoring the Great Lakes ecosystem from 2010 through 2014.[15] The Action Plan was derived from the Strategy, as well as from several other area-specific plans and programs such as the GLWQA and Great Lakes Binational Toxic Strategy,[16] among others. For each focus area under the GLRI, the Action Plan provides a problem statement, a set of goals, interim objectives, progress measures, final targets, and principal activities for restoring the ecosystem. The principal actions in the Plan are not specific projects; rather, they are broad actions that address the objectives of the focus areas. Each year, federal agencies identify projects they plan to take to implement the Action Plan. Project selection is guided by criteria such as:

- ability to achieve measurable outcomes that are linked to high priority issues;
- ability to advance existing Great Lakes activities or the priorities of existing plans for restoring the Great Lakes, including, but not limited to, Lakewide Management Plans, Remedial Action Plans for Areas of Concern,[17] and the Great Lakes Binational Toxic Strategy;[18]
- feasibility of prompt implementation and ability to yield near-term tangible results;
- strong interagency or inter-organizational collaboration and coordination;
- adherence to the best available science;
- public support for the project; and
- low transaction costs and leverage of nonfederal resources.

Projects are geared toward meeting overall and interim goals under each focus area. Overall goals are thematic. For example, Goal 1 under Habitat and Wildlife Protection and Restoration Focus Area states that "protection and restoration of Great Lakes aquatic and terrestrial habitats, including physical, chemical, and biological processes and ecosystem functions, maintain or improve the conditions of native fish and wildlife." Interim goals have quantifiable measures such as "by 2014, 3,000 miles of Great Lakes rivers and tributaries will be reopened and 500 barriers to fish will be removed or bypassed." Measures of progress are quantifiable indicators that are related to the goals. For example, to evaluate progress in restoring habitat and wildlife protection, some measures used are miles of river reopened for fish passage, and percentage of U.S. coastal Great Lakes wetlands assessed. Measures have a quantified baseline and targets for each year between 2010 and 2014.

Monitoring and oversight mechanisms are also discussed in the Action Plan. An accountability system was created to measure and track progress of projects and the implementation of the Action Plan. Further, an interactive electronic map was created that shows the location and description of each project.[19]

Recipients of funding are required to submit status reports on projects and progress towards interim measures and goals of the Action Plan. The EPA collects these reports and compiles an annual report to the President on outcomes and measures.

The report also lists funding allocations for participating agencies by fiscal year. (See *Table 4.)* The most recent report discusses progress in FY2011. There are mixed results for meeting goals and targets, and in several cases, the report states that it is too early to evaluate progress and that several indicators cannot be determined.[20]

Funding

The estimated cost for implementing GLRI for the first five years is approximately $2.2 billion according to the Action Plan. There is no estimate as to how much it will take for the ecosystem to be fully restored, or how long the restoration effort is expected to take. It its inaugural year (FY2010), the GLRI received $475 million in appropriations provided to the EPA.[21] The EPA also was granted the authority to transfer funds to other federal agencies for restoration activities in the Great Lakes and provide grants to state, non-governmental, and private stakeholders. Of the funds appropriated, approximately $400 million were *new* funds that had not been associated with appropriations for existing federal restoration programs. Approximately $75 million was appropriated for existing EPA programs, which addressed Great Lakes restoration.

Funding for the next three years remained steady at approximately $300 million, and the Administration's request for FY2014 is $300 million.[22] (See *Table 1.*) In these appropriations laws, funds were given to the EPA to carry out the GLRI as well as implement activities under the Great Lakes Water Quality Agreement.

Restoration activities are funded through grants and transfers to other federal agencies.

A transfer of funds to an agency is usually done through cooperative or interagency agreements. Funds for other federal agencies are to be considered *new* funds for restoration. Agencies are expected to maintain their base level of funding for restoration activities and identify new activities to support GLRI.

Therefore total annual funding for restoring the Great Lakes ecosystem is funding for the GLRI plus funding for base restoration programs. Base funding for Great Lakes ecosystem restoration activities has been reported in the Great Lakes Ecosystem Restoration Crosscut Budget. For example, for FY2011, base funding for Great Lakes restoration was approximately $677.0 million. With GLRI funding, total funding for restoration was reported at $977.0 million.[23]

Table 1. Funding for GLRI, FY2010-FY2013 and FY2014 Administration's Request (in $millions)

Year	Funding
FY2010	$475.0
FY2011	$300.0
FY2012	$299.5
FY2013	$298.8[a]
FY2014 Request	$300.0

Source: Great Lakes Restoration Initiative at http://greatlakesrestoration.us/index.html and Senate Committee on Appropriations, FY14 Interior Explanatory Statement, at http://www.appropriations.senate.gov/news.cfm? method=news.view&id=d1037190-bf9c-420c-a8a5-79c0ef9c495c.

[a] This figure reflects the 0.2% across the board rescission for FY2013, but not the deductions due to sequestration. Figures are taken from Senate Committee on Appropriations, FY14 Interior Explanatory Statement, at http://www.appropriations.senate.gov/news.cfm?method=news.view&id=d1037190-bf9c-420c-a8a5-79c0ef9c495c.

Implementation

GLRI has received appropriations from FY2010 to FY2013 and is being implemented according to the Action Plan. There has been over a billion dollars appropriated to implement GLRI via its five focus areas through FY2013. (See *Table 2.)* Most of the appropriations have gone to the EPA, with significant amounts of funding going to U.S. Fish and Wildlife Service (FWS), National Oceanic and Atmospheric Administration (NOAA), and the U.S. Army Corps of Engineers (Corps). Note that appropriations for EPA are for EPA directed programs and efforts, as well as grant programs for stakeholder research and restoration activities. (See *Table 3.*)

Table 2. Summary of Funding Allocation by Focus Area, FY2010-FY2013 and FY2014 Administration's Request (in $millions)

Focus Area	FY2010	FY2011	FY2012	FY2013	FY2014 request
Toxic Substances and Area of	$146.9	$100.4	$106.3	n/a	$110.7
Concern					
Invasive Species	60.3	57.5	57.5	n/a	53.0
Nearshore Health and Nonpoint	97.3	49.3	54.8	n/a	56.4
Source Pollution					
Habitat and Wildlife Protection and	105.3	63.0	56.8	n/a	58.8
Restoration					
Accountability, Education,	65.2	29.3	24.1	n/a	21.1
Monitoring, Evaluation,					
Communication, and Partnerships					
TOTAL	475.0	299.4	299.5	298.8	300.0

Source: U.S. Environmental Protection Agency, U.S. Environmental Protection Agency FY2014 Justification of Appropriation Estimates for the Committee on Appropriations, U.S. Environmental Protection Agency, Budget Justification, April 2013, pp. 284-285. n/a = not available.

Note: FY2013 funding for focus area categories has not been reported to include sequestration and the rescission in FY2013.

Table 3. Allocation of GLRI Funds by Federal Agency, FY2010-FY2013 and FY2014 Administration's Request (in $millions)

Agency	FY2010	FY2011	FY2012	FY2013	FY2014 request
U.S. Coast Guard	$6.3	$2.7	$2.7	n/a	$1.9
National Oceanic and Atmospheric	30.5	18.3	15.6	n/a	15.2
Administration					
U.S. Army Corps of Engineers	49.6	31.4	33.8	n/a	20.6
Bureau of Indian Affairs	3.4	6.3	4.7	n/a	4.0
National Park Service	10.5	4.9	3.4	n/a	3.1
U.S. Fish and Wildlife Service	69.3	48.7	43.6	n/a	32.7
U.S. Geological Survey	23.7	14.5	12.4	n/a	11.4
Federal Highway Administration	2.5	1.2	1.2	n/a	1.0
Maritime Administration	4.0	2.7	2.4	n/a	2.3
Agency for Toxic Substances and Disease	5.5	2.2	2.2	n/a	1.7
Registry					
Animal and Plant Health Inspection Service	1.9	0.6	1.1	n/a	0.9
Natural Resources Conservation Service	34.1	16.8	24.2	n/a	23.3
U.S. Forest Service	15.5	8.9	6.7	n/a	6.3
Environmental Protection Agency (including International Joint Commission,	218.1	140.1	145.5	n/a	156.1
Great Lakes Fisheries Commission, and other agreements)					
Multiple Agencies: Asian Carp				n/a	19.5
TOTAL	475.0	299.4	299.5	298.824	300.0

Source: U.S. Environmental Protection Agency, U.S. Environmental Protection Agency FY2014 Justification of Appropriation Estimates for the Committee on Appropriations, U.S. Environmental Protection Agency, Budget Justification, April 2013, pp. 284-285.

Note: FY2013 funding for agencies have not been reported to include sequestration and the rescission in FY2013.

Implementation of GLRI is guided, in part, by provisions in the Interior and Related Agencies Appropriations laws and associated committee and conference reports. In FY2012 appropriations, for example, GLRI was instructed to follow guidelines under H.Rept. 112-151.[25] This committee report stated that EPA should transfer funds to other agencies more expeditiously. Further, the report states that funds are to supplement and expand existing programs, rather than supplant them. The committee report also directed EPA and other federal agencies to prioritize actions that implement "action-oriented" projects in lieu of additional studies, monitoring, and evaluations. Lastly, the report stated that the committee expects to see measurable results from funding over the last few years.[26]

ISSUES FOR CONGRESS

Since FY2010, the implementation of the GLRI represents a significant increase in funding and activities for Great Lakes restoration and a novel attempt at restoring the ecosystem holistically. Some issues with GLRI that have emerged include clarity about the management structure of GLRI, the potential integration of GLRI with existing federal and state restoration efforts in the Great Lakes, the effectiveness of the Action Plan in laying out a strategy for fully restoring the Great Lakes, the overall direction and duration of restoring the Great Lakes, and the funding needed to implement and complete the GLRI. Some of these issues are addressed, in part, by proposed legislation introduced in the 113th Congress. S. 1232 and H.R. 2773 would establish a governance and management structure for restoration activities in the Great Lakes, authorize GLRI and appropriations for its implementation, specify the scope and function of GLRI, and authorize the coordinating role of the Great Lakes Interagency Task Force. A summary of how these bills amend current law and analysis of how they might change current practices in GLRI is provided in the text box below. The rest of this section reviews the aforementioned issues that might be of interest to Congress.

Introduced Legislation Authorizing GLRI in 113th Congress

S. 1232 and H.R. 2773 are two bills in the 113th Congress that would amend the Federal Water Pollution Control Act (33 U.S.C. 1268 (a)) to authorize the GLRI and associated functions and activities related to its implementation.

The bills are largely similar, with some differences in the terms they would define, the structure of meetings held by GLAB, composition of the Task Force, and authorized funding for carry out projects that address remediation of sediment contamination in Areas of Concern. Broadly speaking, both bills would broaden the law to authorize the implementation of GLRI and related activities. Presently, the law authorizes the EPA to conduct restoration activities in the Great Lakes and coordinate activities with other federal agencies.

Both bills would update the findings and purpose of Section 118 to include attaining the goals detailed in the Great Lakes Restoration Initiative Action Plan (Action Plan), Great Lakes Regional Collaboration Strategy (Regional Strategy), and Great Lakes Water Quality Agreement (Agreement). In addition, both bills would expand the forms of federal collaboration in the Great Lakes region to include the funding of contracts and interagency agreements, as well as grants, for the protection, restoration, and pollution control in the ecosystem.

Both bills would amend the law to authorize the Great Lakes Advisory Board (GLAB), the Great Lakes Restoration Initiative (GLRI), and the Great Lakes Interagency Task Force (Task Force). The bills would specifically authorize GLRI to implement projects and activities that would implement the Strategy and Agreement. This would appear to broaden the scope of the GLRI as it would encompass several goals and objectives listed under the Strategy and Agreement. The bills state that federal agencies should maintain funding for baseline activities, and no funds would be provided for water infrastructure projects that also receive funding from state water pollution and drinking water revolving funds.

The Task Force under both bills would be the primary coordinating entity for restoration activities by collaborating with Canada, coordinating the development of federal restoration policies and projects, and assisting in the management of the Great Lakes System. The Task Force would also be responsible for developing outcome-based goals for the ecosystem, and reviewing and updating the Regional Strategy and Action Plan as necessary.

The Task Force would also provide reports to Congress that discusses what actions have been implemented or not implemented with recommendations for changes. The Administrator of the EPA would be responsible for submitting an annual progress report to Congress, and the Director of the Office of Management and Budget would be directed to submit a crosscut budget to Congress.

Both bills would authorize $475 million in appropriations for each of fiscal years 2014 to 2018 to implement GLRI. S. 1232 would authorize $150 million for each fiscal year from 2014 through 2018 for projects determined to address remediation of sediment contamination in areas of concern; H.R. 2773 only would authorize $100 million for each fiscal year from 2014 through 2018 for these projects. In addition, both S. 1232 and H.R. 2773 would authorize an additional $25 million for each fiscal year from 2014 through 2018 for the Great Lakes Program Office.

Implementation and Management of GLRI

Decision-Making

Restoration efforts in the Great Lakes have historically been conducted by several federal or state agencies, largely without a central organizing or governing entity. At the federal level, restoration activities have been conducted under various authorities, with several activities being implemented by the EPA through its Great Lakes National Program Office. In the past, the GAO suggested that a lack of centralized leadership in the Great Lakes restoration activities could detract from the effectiveness and prominence of restoration efforts.[27] Questions such as who is in charge, and how are the implementation of restoration activities to be governed, were posed.[28]

In part, GLRI has addressed these questions in program documents, but some questions about the governance of the restoration initiative remain. For example, it is unclear if a central governing structure that oversees all restoration activities (i.e., all federal, state, and local efforts) exists. The GLRI does not define a central governing structure or decision-making process for guiding all restoration efforts, but appears to promote EPA as the lead federal agency for implementing the initiative. For example, EPA received funds from FY2010 to FY2013 to implement the GLRI and was given authority to transfer these funds to other federal agencies to conduct GLRI activities. EPA was directed to provide annual reports that list funds given to each federal agency and describe program accomplishments. EPA created the Action Plan with the Great Lakes Task Force. While EPA appears to be in charge of implementing GLRI, it does not have the authority to direct the entire Great Lakes restoration effort that includes non-GLRI federal activities and activities done by non-federal stakeholders.[29] The Task Force appears to be responsible for coordinating all restoration activities in the Great Lakes ecosystem and for implementing the Action Plan an integrated manner.[30] These tasks, however, do not appear to constitute decision-making functions;[31] rather it promotes the Task Force as a coordinating entity for all restoration and resource management activities in the Great Lakes.

To temper this issue, the scope of GLRI might not be intended to cover all aspects of Great Lakes restoration. For example, the GLRI is focused on new federal activities and not pre-existing baseline activities, and does not include programs that address infrastructure that affects water quality. However, under this notion, it becomes apparent that GLRI should not be considered a centralized coordinating or management approach to restoring the Great Lakes. This might bring up the earlier question of whether lacking a centralized coordinating or management structure will negatively affect restoration in the Great Lakes.

Coordination

Lacking a coordinated effort to restore the Great Lakes ecosystem is an issue that has been raised in the past for Great Lakes restoration. The GAO asserted that restoration efforts in the Great Lakes suffered from inadequate coordination.[32] GLRI addresses this assertion by stating that it aims to build on existing federal, state, and local activities and draw upon a series of ongoing restoration efforts outlined in existing plans and programs.[33] Further, the Task Force is expected to coordinate Great Lakes restoration efforts among federal agencies.

It is unclear, however, how state and local restoration efforts in the Great Lakes are to be coordinated with the GLRI efforts. Coordination with nonfederal entities in the Great Lakes might be significant due to the number of entities involved in restoration, including efforts from eight states. Some might contend that absent a formal coordination role for nonfederal entities, restoration projects might overlap. Further, some contend that with greater coordination among federal and non-federal stakeholders, there might be opportunities to leverage federal resources with state or local resources to accomplish restoration projects.[34] There is no formal role for nonfederal representatives to coordinate with federal actions in a governance entity. The Task Force does not have nonfederal members. Coordination among nonfederal members could be addressed by the GLAB, however their charter does not specify that coordination is a priority. GLAB is to provide advice on the implementation of GLRI. Some have also questioned whether GLRI should have a formal collaboration with Canada and Ontario to coordinate restoration activities.

Some restoration activities conducted by nonfederal entities are done under the GLRI with grants from federal agencies. Grants are awarded for projects in one of the focus areas and grantees are required to input project information (description of work and project outputs) into the Great Lakes Accountability System (GLAS). This information could be used to facilitate coordinated approaches to restoration and to determine if projects have similar objectives, but non-GLRI projects are not inserted into GLAS making analysis of overlap or potential for collaboration difficult.

Outside of GLRI efforts there have been attempts to organize and report all federal and state activities related to ecosystem restoration in the Great Lakes. For example, in the last few years, the creation of a Great Lakes crosscut budget has been mandated by Congress in annual appropriations laws.[35] The resulting crosscut budget contained information on GLRI funding as well as other base federal funding for Great Lakes restoration. It also reported some state spending for restoration efforts; however, just Wisconsin reported funding for FY2012. Some might view the creation and dissemination of a broad crosscut budget as one tool to organize and potentially coordinate all Great Lakes restoration activities.

Examples from Other Restoration Initiatives

There are several approaches for addressing governance and coordination that are exemplified in other large-scale restoration initiatives that could be contemplated for the GLRI. For example, Congress could create a single entity that centrally coordinates restoration activities at all stakeholder levels. This could include all federal activities and state activities. The restoration initiative in the Platte River is guided by a Governance Committee (GC), which is responsible for implementing the Program.[36] The GC is a ten-member body with representatives from Colorado, Nebraska, and Wyoming, U.S. Bureau of Reclamation (USBR), U.S. Fish and Wildlife Service (FWS), water users, and environmental entities. The GC makes programmatic decisions including changes to budgets and changes to restoration activities. The GC justifies these types of changes with new information or recommendations from adaptive management efforts. An Oversight Committee, consisting of the Secretary of the Interior and the Governors of Colorado, Nebraska, and Wyoming, is responsible for approving significant modifications to the program.[37]

Several other ecosystem restoration initiatives are governed by hybrid entities with both federal and state partners. In some cases, these governing bodies are limited in their ability to make decisions. Their decision-making authority extends to certain types of decisions such as creating a restoration plan, conducting science, and monitoring restoration projects. However, project implementation might be left to individual state and federal agencies rather than the hybrid entity. Examples of this type of governing entity include the Chesapeake Bay Executive Council, Long Island Sound Study Policy Committee, and the South Florida Ecosystem Restoration Task Force.

S. 1232 and H.R. 2773

Bills in the 113th Congress address governance and coordination under GLRI. S. 1232 and H.R. 2773 would authorize the EPA to select programs and projects for Great Lakes protection and restoration. The bill states that the EPA would consult with federal partners, including the Task Force, and consider recommendations of GLAB, when selecting projects to implement.[38] The bill would also authorize the Task Force and direct it to carry out several coordinating activities related to restoring the Great Lakes ecosystem. Some examples include collaborating with Canada and bi-national bodies involved in activities related to the Great Lakes; managing the Great Lakes system, considering recommendations from GLAB; coordinating government actions associated with implementing restoration plans such as the Action Plan, Strategy, and GLWQA; and developing outcome-based goals for the Great Lakes System.[39] Under these bills, EPA would be the lead agency in implementing GLRI, and the Task Force would be the lead entity in coordinating overall federal and nonfederal efforts to address the Great Lakes ecosystem.

In addition, these bills would require the GLRI to prioritize work done by nonfederal partners for priority areas each year and would require that federal projects and nonfederal projects are implemented in coordination with states and other organizations.[40] Further, the bills would direct the Office of Management and Budget to create a Great Lakes crosscut budget to report funding and projects funded by GLRI and potentially other non-GLRI activities. The crosscut budget would also be required to identify all expenditures by federal and state governments on Great Lakes restoration activities since FY2004. If enacted, these bills would create a permanent requirement for creating a crosscut budget.

GLRI Action Plan

The lack of a comprehensive plan or strategy to guide restoration efforts in the Great Lakes was a concern in the past for restoration efforts in the Great

Lakes.[41] Without a plan, the GAO reported that organizations developed their own strategies for restoration, inadvertently making coordination among them difficult. The GLRI Action Plan was created in part to address this concern. The Action Plan discusses broad themes related to restoration and provides a set of short term goals for measuring progress. The Action Plan specifically states that it does not address water infrastructure programs that include the Clean Water or Drinking Water State Revolving Fund program, nor base federal agency restoration activities.

Some could contend that the Action Plan still is not a comprehensive plan for restoration because some restoration themes and projects are not included. For example, the Action Plan lists themes that are to be addressed by restoration activities, but does not provide details on specific restoration projects that are to be implemented or how projects are connected to the restoration of the ecosystem. Further, the Action Plan does not cover all federal restoration activities (e.g., those ongoing restoration activities initiated before GLRI and the Action Plan were created) and does not include or address all state or local restoration activities.

To counter these points some might contend that the scope of the Action Plan is only based on implementing GLRI (i.e., not all federal restoration activities) and that specificity of projects to be funded and implemented depends on annual decisions and recommendations made by the EPA, participating federal agencies, and the Task Force. Therefore, it would not be possible to list all restoration projects upfront. Further, they might contend that broad restoration actions in the Great Lakes ecosystems are also, in part, guided by the Strategy and the Great Lakes Water Quality Agreement. The Strategy lists thematic areas for restoration and key actions for restoration; and GLWQA contains long term goals for restoration that would encompass all restoration activities related to water quality. Both bills in the 113th Congress would authorize restoration actions to achieve goals established in the Action Plan, Strategy, and GLWQA.[42] This would create a broader restoration effort that goes beyond just implementing the Action Plan and would include actions and goals in both the Strategy and GLWQA.

Issues related to the implementation of the Action Plan could raise questions about how the Action Plan will integrate GLRI activities with existing federal, state, and local activities. The GLRI does not include all federal and nonfederal actions related to restoring the Great Lakes in the Action Plan.[43] It specifically states that it will guide the implementation of the GLRI, not necessarily all restoration efforts in the Great Lakes ecosystem.[44] The Action Plan does state that it will build upon past programs and plans that are now in place. It is unclear if the project selection, oversight, and reporting requirements under GLRI directly cover non-GLRI activities in addition to taking them into consideration.

Examples of Non-GLRI Activities Related to Restoration

There are several restoration activities in the Great Lakes that are not covered under GLRI. These activities are termed base-level restoration activities, and are generally restoration activities that federal agencies have been implementing in the Great Lakes ecosystem before GLRI was initiated. The exception to this description is some base activities implemented by the EPA that directly address the Great Lakes (e.g., remediation of contaminated sediments in the Great Lakes ecosystem), which are included in GLRI. Further, several of the base-level activities are implemented on a national level with a national level scope and purpose. A portion of program funds are invested in the Great Lakes region and these activities generally have secondary benefits for restoring the Great Lakes ecosystem. The appropriations estimated for non-GLRI restoration activities are much higher than GLRI appropriations. For example, in FY2012, non-GLRI restoration activities were appropriated approximately $579 million (GLRI was appropriated $300 million). Of this amount, nearly 87% was for national level programs.

Some selected examples of non-GLRI restoration activities include:

- Agricultural conservation programs administered by the Natural Resources Conservation Service such as the Environmental Quality Incentives Program and the Wetlands Reserve Program
- Clean Water State Revolving Fund administered by the EPA
- Coastal Zone Management grants administered by the National Oceanic and Atmospheric Administration
- National Wildlife Refuge System (in the Great Lakes region) and Great Lakes Fish and Wildlife Restoration program administered by FWS
- Great Lakes Fishery Commission overseen by the U.S. Department of State

If GLRI does not cover all new and existing federal activities, some questions might arise, including:

- Will GLRI reporting and oversight activities include non-GLRI funded efforts?
- How will GLRI integrate non-GLRI projects in the restoration effort, or how would duplication of efforts be avoided?
- Are interim goals and overall goals a true indication of GRLI efforts?
- Will federal agencies maintain non-GLRI activities or eventually integrate them into GLRI?

Some of these questions might be addressed in the pending revision to the Action Plan which addresses restoration activities from 2015 to 2019. Some contend that the revised Action Plan should contain greater emphasis on monitoring and adaptive management and focus on achieving and assessing outcome-oriented results (i.e., results that reflect the condition of the ecosystem) rather than just output-oriented results (i.e., results that are reported from specific projects).[45] Creating a revised Action Plan based on the results and outcomes of projects implemented under the first Action Plan could be difficult if sufficient results have not been reported in time for the revision. This might not allow for the revised Action Plan to incorporate lessons learned and project results from the first phase of restoration.

Vision of a Restored Great Lakes Ecosystem

The GLRI does not present an overall vision for what a restored Great Lakes ecosystem should look like and function. An overall vision is broadly discussed in the Action Plan,[46] but specific goals and targets for restoration are provided for just the first four years of restoration. Furthermore, there is no indication of how long it will take to restore the Great Lakes ecosystem. The Task Force, under GLAB, is currently revising the Action Plan to describe activities and goals from FY2015 to FY2019, yet it is unclear if this represents a stage for restoration or a plan to complete restoration. Some might counter this concern by noting that it is difficult to fully contemplate a restored Great Lakes ecosystem because of its size and complexity. They might also point to the Great Lakes Water Quality Agreement as a guiding document for long-term restoration. The Agreement contains general objectives for restoring the waters of the Great Lakes, including that the waters should be a source of high-quality drinking water, free from pollutants that could be harmful to humans, and support healthy and productive wetlands, among other things.[47] Further, the Agreement contains 10 Annexes that describe long-term objectives for addressing specific issues related to restoration such as diminishing excess nutrient deposition (e.g., phosphorus and nitrogen) and controlling aquatic invasive species, among others. In the GLWQA there is no indication of how long it will take to restore the waters of the Great Lakes, nor estimate of how much it will cost.

The lack of defining what a restored ecosystem might resemble under the GLRI or how long it might take restore the ecosystem could generate questions related to how much restoring and maintaining the Great Lakes ecosystem could ultimately cost. Other questions might include what the long-terms plans are for restoration and if any major restoration projects are being contemplated; and if restoration might involve manipulating water flows into or out of the Great Lakes to enhance habitat for native species or restrict the entry of aquatic nuisance species. Answers to these questions could involve actions that could be potentially controversial. [48]

Long-term issues associated with the Great Lakes ecosystem such as the control and eradication of invasive species or the potential effects of climate change might not be adequately addressed in five-year increments. This and other restoration issues might require a longer planning horizon than five years to address. The potential effects of climate change on the Great Lakes ecosystem should be incorporated into the Action Plan according to some.[49] The Action Plan addresses climate change by noting that it could have implications for focus areas and would be addressed where appropriate. Some contend that addressing the effects of climate change in the selection and implementation of restoration projects is important for the long term viability of restoration.[50] Climate change could potentially alter Great Lakes water levels, which have a significant connection to several ecosystem properties and economic factors in the Great Lakes.

S. 1232 and H.R. 2773 address this issue, in part, by authorizing the Task Force to review the Strategy and Action Plan every five years. The Task Force would be authorized to update and revise the Action Plan. This policy would allow for revisions to reflect changes in the ecosystem, but might not generate a vision for the restored ecosystem or result in providing the estimated duration of restoration.

Progress of Restoring the Great Lakes under the GLRI

Implementation of GLRI began in 2010; however, it is unclear how much more is needed to restore the ecosystem. GLRI aims to track progress of restoration within the timeframe of the Action Plan. GLRI has an accountability system (Great Lakes Accountability System; GLAS) that lists projects and provides a description and their funding.[51] Further, GLAS requests project managers to identify the focus area they are working under and any quantifiable results from the project (e.g., acres of wetlands restored). GLAS documentation does suggest that results from projects could be aggregated and tracked, and that information will allow managers to adjust priorities based on data derived from GLAS. However, based on this description, some might contend that it is unclear how the completion or progress of these projects relates to the overall restoration of the ecosystem or ecosystem processes. For example, one researcher commented that many of the restoration projects are completed in "silos" without a comprehensive assessment of how projects relate to overall restoration.[52] In addition, he questioned whether measuring and reporting indicators will provide an understanding of how the overall ecosystem is improving or increasing its resilience. Others emphasize that outcomes of the restoration initiative are unclear, even though indicators and measurements are reported. In response to these claims, GLRI notes that monitoring and evaluation of progress is conducted on a regular basis.[53] Further, GLAS is expected to release a set of "dashboard indicators" that aim to provide a visual representation of progress toward metrics.[54] To address this point, some have called for broadening monitoring, and reframing the analyses and reporting of the results so that progress toward outcomes can be evaluated.[55] Both proposed bills authorizing GLRI in the 113th Congress would direct the Task Force to create outcome based goals for the Great Lakes. This could address, in part, the issue of evaluating restoration progress at the regional or ecosystem level, rather than at the project level. However, developing outcome based goals could be perceived as different from developing a strategy for achieving outcomes.

The GLRI accountability approach under GLAS appears only to be applicable to GLRI projects. It is unclear if it will be expanded to include non-GLRI and nonfederal activities. Without taking into account all efforts to restore the ecosystem, it would be difficult to evaluate the effectiveness and progress of GLRI. If GLRI and non-GLRI activities are contributing to restoration, how does one evaluate only GLRI activities? Progress toward meeting goals was reported in annual reports to Congress for FY2010 and FY2011. Progress reported for GLRI is mixed in these reports with several entries stating that indicators have not yet been determined. The reports also do not differentiate between progress made by GLRI projects versus non-GLRI federal and state actions. To counter this sentiment, one could contend that progress toward goals may not be measurable after a few years of implementing restoration activities. For many ecological indicators, progress might not be seen until several years after restoration actions have been implemented.

There do not appear to be any consequences for not progressing or reaching goals set by the GLRI. For example, if restoration targets are not being met, there does not appear to be a policy mechanism to alter the implementation or direction of GLRI efforts. Several other restoration initiatives have had mixed experiences working with quantitative goals and indicators of progress. For example, in the Chesapeake Bay and the Everglades restoration initiatives, some initial goals have not been met on schedule, resulting in criticism of the initiative. Policy mechanisms to address these shortcomings include revising the goals and indicators of progress, or implementing an adaptive management program.[56] Adaptive management can also be a policy tool that would allow changes in projects and goals if new circumstances arise or if progress stalls. The GLRI encourages agencies to incorporate adaptive management in implementing restoration projects, but has not implemented an adaptive management process. A recent draft report, however, proposes an adaptive management framework for restoration activities in the Great Lakes.[57] The objective of this framework would be to address programmatic decisions related to restoration. It would also address priorities for implementing projects and refine decision-making. Adaptive management can also be implemented on the programmatic scale. It could be used to measure and potentially adjust how the program is being implemented and how overall restoration objectives are being met.[58]

The lack of long-term restoration targets and a vision for a restored ecosystem might make measuring the progress of restoration difficult. Indeed, some have noted that GLRI projects and their outputs should be connected to an over-arching plan for restoration.[59] GLRI has a set of defined targets that are intended to be a measure of progress for restoring the Great Lakes. The targets are quantifiable and are listed for each year between 2010 and 2014 under each focus area.[60] However, restoration targets beyond the five-year frame are not discussed in the Action Plan.[61] This brings up the question of how many goals or steps are needed to be met for fully restoring the Great Lakes, and where is the current Great Lakes ecosystem on this scale. Several other large-scale ecosystem restoration initiatives have estimated the total duration and cost for restoration, yet most of them have had to alter their funding and time of completion estimates as restoration progressed.

Value of Restoring the Great Lakes

The question of whether an ecosystem restoration initiative is worth funding and whether restoration will ultimately provide an overall net financial benefit to the region has been brought up for several ecosystem restoration initiatives, including the Great Lakes. Answering this question for the Great Lakes is complicated since the total cost of restoration and its duration is unclear. The Brookings Institution has asserted that restoring the Great Lakes will not only help the ecosystem, but will provide an economic benefit to the region.[62] Their 2007 study reported the estimated financial benefits associated with improvements in Great Lakes environmental quality included higher catch rates in recreational fisheries, lower water treatment costs, more and better swimming opportunities, improved bird watching, more hunting opportunities, and benefits from removal of contaminated sediment. The sum of present-value benefits in 2007 from specific improvements in the environment totaled approximately $50 billion for long-term gains and $30-$50 billion for short-term gains.[63] The full cost of restoration, however, was not identified in this study making it impossible to evaluate the net benefit of restoration.

The total cost of funding the complete restoration of the Great Lakes ecosystem has not been estimated by the GLRI. GLRI estimates that $2.2 billion in funding is needed for the first five years of the restoration. It is unknown how much funding it will take to restore the ecosystem and what the ultimate benefits of restoration might be.[64] Further, it is unclear where along the restoration timeline GLRI might be because the problem and proposed solution has not been fully defined. Reporting by GLRI does not project how much longer restoration might take. Whether the first phase (2010 through 2014) covers a small or large space along the restoration timeline might influence decisions on how much funding to provide for the initiative.

Questions related to the cost of restoring the Great Lakes ecosystem and whether that cost is justified could be heightened because annual funding for restoring the Great Lakes is larger than that for other prominent ecosystems. Funding for all federal restoration activities in the Great Lakes has ranged from $649 million to $687 million annually from FY2004 to FY2009.[65] With GLRI, the FY2010 federal funding for Great Lakes restoration was approximately $1.2 billion (non-GLRI funding was approximately $720 million). Other ecosystem restoration efforts in areas such as the Everglades and Chesapeake Bay have lower annual federal funding (e.g., approximately $350 million and $400 - $450 million annually, respectively). Some might contend that the expansive size of the Great Lakes ecosystem warrants a higher price tag than other large-scale ecosystem restoration initiatives in the United States, and that funding needs for restoration are on par with other ecosystems. Further, they might argue that the Great Lakes ecosystem is one of the largest in the country, and potentially could contribute the greatest value to people. However, some others might contend that federal funding for GLRI is too high because combined with non-GLRI funding it is higher than any other ecosystem.

S. 1232 would authorize $475 million for each fiscal year from FY2014 to FY2018 to carry out GLRI. Further, it would authorize $150 million for each fiscal year during the same period to carry out projects that address remediation of sediment contamination in Areas of Concern, and $25 million for each fiscal year during the same period for the Great Lakes Program Office to administer and conduct restoration activities. H.R. 2773 would authorize the same amount of funding for GLRI and the Program Office, and $100 million for addressing remediation of sediment contamination in Areas of Concern.

CONLCUSION

The GLRI is a large-scale ecosystem restoration initiative that aims to restore the Great Lakes ecosystem under the framework of an Action Plan. It attempts to address past criticisms of Great Lakes restoration activities that included claims of a loosely coordinated restoration effort, and a restoration effort that lacks a comprehensive plan. The Great Lakes ecosystem is complex and covers a vast geographical area. This inherently leads to scientific uncertainties in implementing restoration projects and unforeseen circumstances in the evolution of the ecosystem. These factors and others may cause restoration efforts under the GLRI to face persistent and new questions by Congress and other stakeholders. Some might question how long restoration in the Great Lakes may take and how much it could ultimately cost. Further, questions related to how the GLRI is implemented and whether it will eventually encompass all federal activities, state activities, and local activities related to Great Lakes restoration could be asked and discussed. The GLRI is not authorized in law, but in the 113th Congress, two bills were introduced to authorize the initiative. Both bills would address similar issues, including authorizing the EPA as the lead agency for implementing GLRI, and authorizing the Task Force to coordinate among federal and non-federal stakeholders. They would also authorize $475 million annually from FY2014 to FY2018 to fund GLRI activities.

End Notes

[1] These are species listed under the U.S. Endangered Species Act (16 U.S.C. §§1531-1544). For a more detailed summary of causes and symptoms of ecosystem deterioration in the Great Lakes, see Donald Scavia et al., Prescription for Great Lakes Ecosystem Protection and Restoration, National Wildlife Federation, Report, December 2005, pp. 1- 39.

[2] For more information on U.S. federal and state programs aimed at restoring the Great Lakes, see U.S. Government Accountability Office, Great Lakes: An Overall Strategy and Indicators for Measuring Progress Are Needed to Better Achieve Restoration Goals, GAO-03-515 (Washington, DC: April 2003). Hereafter referred to as An Overall Strategy and Indicators for Measuring Progress Are Needed to Better Achieve Restoration Goals.

[3] John C. Austin et al., Healthy Waters, Strong Economy: The Economic Benefits of Restoring the Great Lakes Ecosystem, The Brookings Institution, September 2007.

[4] For example, see An Overall Strategy and Indicators for Measuring Progress Are Needed to Better Achieve Restoration Goals, and Melissa Malott, Restoration Plan Key to Helping Lake Michigan, Other Great Lakes, Clean Wisconsin, May 29, 2008.

[5] The Agreement addresses shared priorities of the United States and Canada with respect to the Great Lakes and aims to restore and protect the chemical, physical, and biological integrity of the waters of the Great Lakes.

[6] See An Overall Strategy and Indicators for Measuring Progress Are Needed to Better Achieve Restoration Goals.

[7] Ibid.

[8] The Great Lakes Interagency Task Force was created by an executive order in 2002 (E. O. 12240). Its purpose is to provide strategic direction on federal Great Lakes policy. The Task Force is chaired by the Administrator of the U.S. Environmental Protection Agency (EPA) and contains 10 agency and cabinet-level officers.

[9] The Great Lakes Regional Collaboration Strategy can be found at http://www.glrc.us/strategy.html.

[10] GLRI does not include water supply issues, which are addressed through the Great Lakes-St. Lawrence River Basin Water Resources Compact. Water infrastructure needs in the Great Lakes are not addressed by the GLRI, but through funding provided by the Drinking Water State Revolving Loans and the Clean Water State Revolving Funds in EPA.

[11] Areas of Concern are geographic areas within the Great Lakes Basin that have been designated for restoration under Annex 2 of the Great Lakes Water Quality Agreement. Specifically, these areas " fail to meet the general or specific objectives of the Agreement where such failure has caused or is likely to cause impairment of beneficial use of the area's ability to support aquatic life."

[12] U.S. Environmental Protection Agency Charter, Great Lakes Advisory Board, Filed June 22, 2012, http://www.epa.gov/ocem/faca/pdf/2012/2012_glab_charter_establishment.pdf.

[13] Ibid, section 3.

[14] White House Council on Environmental Quality, U.S. Environmental Protection Agency, et al., Great Lakes Restoration Initiative Action Plan, Great Lakes Interagency Task Force, February 21, 2010. Hereafter referred to as The GLRI Action Plan.

[15] The Task Force is planning to revise the Action Plan for FY2015 to FY2019.

[16] The Great Lakes Binational Toxic Strategy is a strategy put forth by Canada and the United States that aims to virtually eliminate persistent toxic substances resulting from human activities in the Great Lakes Basin. See the Strategy at http://binational.net/bns/strategy_en.pdf

[17] In 1987, the United States and Canada identified 43 Areas of Concern (AOC) in the Great Lakes basin that represented the most degraded portions of the ecosystem. The most common reason for degradation in AOCs is contaminated sediments.

[18] For a full list of pre-existing plans, see page 11 of The GLRI Action Plan.

[19] See http://greatlakesrestoration.us/

[20] U.S. Environmental Protection Agency and Great Lakes Interagency Task Force, Great Lakes Restoration Initiative: Fiscal Year. 2011 Report to Congress and the President, U.S. Environmental Protection Agency, September 2011, pp. 1-44, http://greatlakesrestoration.us/pdfs/2011-glri-report-to-congress.pdf.

[21] P.L. 111-88.

[22] FY2013 funding numbers are not necessarily comparable to other years because post-sequestration funding estimates are not publically available.

[23] Office of Management and Budget, Great Lakes Crosscut Report to Congress, Office of Management and Budget, May 2011, pp. 38.

[24] Figures are taken from Senate Committee on Appropriations, FY14 Interior Explanatory Statement, at http://www.appropriations.senate.gov/news.cfm?method=news.view&id=d1037190-bf9c-420c-a8a5-79c0ef9c495c.

[25] U.S. Congress, House Committee on Appropriations, Department of Interior, Environment, and Related Agencies Appropriations Bill, 2012, Report to Accompany H.R. 2584, 112th Cong., 1st sess., July 19, 2011, H.Rept. 112-151 (Washington: GPO, 2011), pp. 64-65.

[26] U.S. Congress, House Committee on Appropriations, Department of Interior, Environment, and Related Agencies Appropriations Bill, 2012, Report to Accompany H.R. 2584, 112th Cong., 1st sess., July 19, 2011, H.Rept. 112-151 (Washington: GPO, 2011), pp. 64-65.

[27] U.S. Government Accountability Office, Organizational Leadership and Restoration Goals Need to Be Better Defined for Monitoring Restoration Progress, Highlights of GAO-04-1024, September 2004, p. 1.

[28] Ibid.

[29] Non-federal stakeholders are expected to receive funding from GLRI through grants. They are also expected to work with EPA to establish a process to provide guidance for implementing the GLRI. Further, one longterm goal of the draft Action Plan is to create mechanisms that give stakeholders and citizens the opportunity to provide input to governments on Great Lakes issues and concerns.

[30] GLRI Action Plan, p. 13.

[31] Decision-making actions could include actions such as directing which projects to implement; directing the implementation of the restoration initiative, including making changes to implementation; defining the scope of restoration; and allocating funds.

[32] An Overall Strategy and Indicators for Measuring Progress Are Needed to Better Achieve Restoration Goals, p. 35.

[33] GLRI Action Plan, p. 11.

[34] For example, see Lynn McClure, Joel Brammeier, and John Jackson, Comments on the Great Lakes Restoration Initiative Action Plan Update, Healing Our Waters-Great Lakes Coalition, July 12, 2013, p. 7. Hereafter referred to as Comments on the GLRI Action Plan Update.

[35] For example, see P.L. 112-74, section 737.

[36] For more information, see https://www.platteriverprogram.org/AboutPRRIP/Pages/Program Information.aspx.

[37] The modifications are specified in the Final Platte River Recovery Implementation Program, and include such factors as changing the regulatory certainty afforded under the Program.

[38] Section 2(b)(4) of S. 1232.

[39] Section 2(b)(5) of S. 1232.

[40] Section 2(b)(4) of S. 1232.

[41] An Overall Strategy and Indicators for Measuring Progress Are Needed to Better Achieve Restoration Goals, p. 35.

[42] See section 2(a)(2) of S. 1232.

[43] The GLRI Action Plan, p. 13.

[44] The Action Plan notes that it is not the "only tool in the toolbox" referring to other programs that address restoration. See The GLRI Action Plan, p. 4.

[45] For example, see Comments on the GLRI Action Plan Update, p. 2.

[46] The GLRI Action Plan, p. 9.

[47] Article 3(1)(a) of the Great Lakes Water Quality Agreement.

[48] This issue has generated controversy among stakeholders bordering Lake Ontario. A proposal by the International Joint Commission to alter water flows from the Moses Saunders Dam and other structures for ecosystem restoration and other purposes have caused some property owners along the southern shore of Lake Ontario to criticize the plan because of its potential to flood their property.

[49] Comments on the GLRI Action Plan Update, p. 3.

[50] Comments on the GLRI Action Plan Update, p. 3.

[51] U.S. Environmental Protection Agency, Great Lakes Restoration Initiative Accountability System User Guide, Version 1.11, March 2012, p. 1, http://www.glri.us/granteeinfo.html.

[52] Don Scavia, "Avoiding the Tipping Point: A Decade of Developments," Session Talk, Milwaukee, WI, September 10, 2013.

[53] Cameron Davis, "Avoiding the Tipping Point: A Decade of Developments," Response to Session Talk, Milwaukee, WI, September 10, 2013.

[54] U.S. Environmental Protection Agency, Great Lakes Restoration Initiative Accountability System User Guide, Version 1.11, March 2012, p. 1, http://www.glri.us/granteeinfo.html.

[55] Don Scavia, "Avoiding the Tipping Point: A Decade of Developments," Session Talk, Milwaukee, WI, September 10, 2013.

[56] Adaptive management is the process of incorporating new scientific and programmatic information into the implementation of a project or plan to ensure that the goals of the activity are being reached efficiently. It promotes flexible decision-making to modify existing activities or create new activities if new circumstances arise (e.g., new scientific information) or if projects are not meeting their goals.

[57] Science Subgroup of the Great Lakes Regional Working Group, Great Lakes Restoration Initiative: Adaptive Science-Based Framework for Great Lakes Restoration, Great Lakes Regional Working Group, Draft Report, May 21, 2013, pp. 1-32.

[58] For more information on adaptive management, see CRS Report R41671, Adaptive Management for Ecosystem Restoration: Analysis and Issues for Congress, coordinated by Charles V. Stern.

[59] Comments on the GLRI Action Plan Update, p. 7.

[60] These measures of progress are listed in the Action Plan.

[61] The Great Lakes Water Quality Agreement, however, has some long-term restoration targets and objectives that do not appear to be time-limited.

[62] John C. Austin et al., Healthy Waters, Strong Economy: The Economic Benefits of Restoring the Great Lakes Ecosystem, The Brookings Institution, September 2007.

[63] John C. Austin et al., Healthy Waters, Strong Economy: The Economic Benefits of Restoring the Great Lakes Ecosystem, The Brookings Institution, September 2007.

[64] For comparison, the Great Lakes Regional Collaboration Strategy calls for spending $20 billion over a five-year period to restore the Great Lakes ecosystem. Of this total, $13.7 billion in new funds is for wastewater treatment system improvements (the anticipated federal share is $7.5 billion and nonfederal share is $6.2 billion), and $1.3 billion for improvement of drinking water treatment facilities. These activities are specifically not included in the Action Plan. See The Great Lakes Regional Collaboration, The Great Lakes Regional Collaboration Strategy (December 2005), accessed at http://www.glrc.us/.

[65] Office of Management and Budget, Great Lakes Restoration Crosscut: Report to Congress, Office of Management and Budget, Washington, DC, March 2010, http://www.whitehouse.gov/sites/default/files/omb/assets/ legislative_reports/2010_great_lakes_report.pdf.

In: The Great Lakes Restoration Initiative
Editor: Óskar Strobel
ISBN: 978-1-63117-525-1

Chapter 2

GREAT LAKES RESTORATION INITIATIVE: FURTHER ACTIONS WOULD RESULT IN MORE USEFUL ASSESSMENTS AND HELP ADDRESS FACTORS THAT LIMIT PROGRESS[*]

United States Government Accountability Office

WHY GAO DID THIS STUDY

The Great Lakes contain about 84 percent of North America's surface freshwater and provide economic and recreational benefits in the Great Lakes Basin. However, the Great Lakes face significant stresses—such as toxic pollution—that have caused ecological and economic damage to the region.

Approximately $1.3 billion has been appropriated to the GLRI, created in fiscal year 2010, which an interagency Task Force of 11 federal agencies, chaired by the EPA Administrator, oversees. In 2010, the Task Force issued an Action Plan for fiscal years 2010 to 2014 to develop a comprehensive approach to restoring the health of the Great Lakes ecosystem. GAO was asked to review the GLRI. This report examines (1) how the GLRI is implemented by the Task Force agencies and other stakeholders, (2) the methods that EPA has in place to assess GLRI progress, (3) the progress

[*] This is an edited, reformatted and augmented version of the United States Government Accountability Office Publication, No. GAO-13-797, dated September 2013.

identified by the Task Force agencies and nonfederal stakeholders, and (4) the views of nonfederal stakeholders on factors, if any, that may affect or limit GLRI progress. GAO analyzed the Action Plan, surveyed 205 non-federal recipients of GLRI funding, and interviewed Task Force agency officials and nonfederal stakeholders.

WHAT GAO RECOMMENDS

GAO recommends that EPA help ensure more comprehensive and useful GLRI progress assessments and account for factors outside of the Action Plan's scope that may affect the GLRI's long-term success. EPA generally agreed with GAO's recommendations.

WHAT GAO FOUND

The Task Force agencies use the Action Plan to implement the Great Lakes Restoration Initiative (GLRI) and use an interagency process to enter into agreements among themselves to identify GLRI projects and with other stakeholders to implement GLRI projects. The Action Plan includes guidance for implementing the GLRI in five focus areas (such as invasive species and habitat and wildlife protection and restoration) that encompass the most significant environmental problems in the Great Lakes. Each focus area includes, among other things, long-term goals, objectives to be achieved by fiscal year 2014, and 28 measures of progress that have annual targets for fiscal years 2010 to 2014.

EPA uses the Action Plan's measures to assess GLRI progress. However, its methods may not produce comprehensive and useful assessments of GLRI progress for several reasons. Among them, some of the goals and objectives do not link to any measures and, as a result, it is unclear how EPA will be able to assess progress toward them. In addition, some measures track actions that may not lead to the desired GLRI goal. For example, one measure tracks the reduction in concentrations of polychlorinated biphenyls (PCB) in fish as part of the goal to lift all restrictions on consumption of Great Lakes fish. However, stakeholders reported that the measure is too narrow and that mercury and other contaminants need to be addressed as well. Consequently, reducing PCB concentrations in fish is not likely to lead to the desired result of lifting all Great Lakes fish consumption restrictions. Without useful measures, EPA may not be able to determine that GLRI efforts are producing the desired results.

In spring 2013, the Task Force agencies issued two reports about GLRI progress in fiscal years 2010 and 2011, which state whether the targets for the Action Plan's 28 measures were being met (e.g., 15 of 28 measures met or exceeded in fiscal year 2011), but the reports include few specific examples of progress. As a result, GAO sought further insights into such progress by surveying nonfederal GLRI stakeholders. Overall, 87 percent of respondents cited at least one example of how one or more of their projects had, or was expected to, benefit the Great Lakes ecosystem. GAO and others have reported that quantifying overall Great Lakes restoration progress is difficult, that the environmental conditions of each lake are unique, and, according to a 2006 U.N. report, it is often impossible to attribute specific environmental changes to specific projects or programs.

In response to GAO's survey, among the factors respondents most often cited as potentially limiting GLRI progress are several outside the scope of the Action Plan, such as inadequate infrastructure for wastewater or stormwater and the effects of climate change. These factors could negatively affect GLRI restoration efforts. For example, as a result of climate change, warming water temperatures can lead to increased numbers of aquatic invasive species and a decline in native ones, a GLRI focus area. The Action Plan touches on these factors but does not state how they will be addressed. In 2012, EPA took steps to incorporate climate change considerations into a small number of GLRI projects but has yet to decide if the GLRI will consider climate change impacts on all GLRI projects. Without addressing these factors in the next Action Plan, EPA will not be able to more fully account for their impacts on GLRI restoration efforts.

ABBREVIATIONS

Action Plan	Great Lakes Restoration Initiative Action Plan
EPA	Environmental Protection Agency
GLAS	Great Lakes Accountability System
GLRI	Great Lakes Restoration Initiative
GPRA	Government Performance and Results Act of 1993
PCB	polychlorinated biphenyls
Task Force	Great Lakes Interagency Task Force

441 G St. N.W.
Washington, DC 20548

September 27, 2013

The Honorable Bill Shuster
Chairman
Committee on Transportation and Infrastructure
House of Representatives

The Honorable Bob Gibbs
Chairman
Subcommittee on Water Resources and Environment
Committee on Transportation and Infrastructure
House of Representatives

The Great Lakes form the largest system of freshwater on earth. They contain about 84 percent of North America's surface freshwater and are shared by eight U.S. states—Illinois, Indiana, Michigan, Minnesota, New York, Ohio, Pennsylvania, and Wisconsin—the Canadian province of Ontario, and more than 40 tribes. The Great Lakes provide a variety of recreational, environmental, and economic benefits, including drinking water, fishing, swimming, boating, agriculture, industry, and shipping, for the more than 30 million people who reside in the Great Lakes Basin, which includes the five Great Lakes and the large land area that extends beyond the lakes, including their watersheds, tributaries, and connecting channels. In addition, a 2011 Michigan Sea Grant study estimated that 1.5 million jobs in the United States were directly connected to the Great Lakes and generated $62 billion in wages in 2009.[1]

However, the Great Lakes face significant environmental and public health stresses that continue to threaten their future. For example, advisories that warn against the consumption of certain fish species have been issued for each of the Great Lakes states due to their levels of persistent toxic substances, such as mercury. Decades of industrial activity in the region have left a legacy of contamination, such as from polychlorinated biphenyls (PCB), that resulted in the United States and Canada identifying, since 1987, a list of 43 severely degraded locations in the Great Lakes Basin as areas of concern; 26 of these areas of concern are located wholly in the United States, and 5 are shared between the two nations. In addition, more than 180 nonnative or "invasive" aquatic species have become established in the Great Lakes. Some invasive species have caused extensive ecological and economic damage to the Great Lakes. The zebra mussel, for example, outcompetes native species for resources by removing live animals and algae that other fish eat from water bodies, and it blocks pipes that deliver drinking water to cities and cooling water to power plants. The Fish and Wildlife Service has estimated a potential economic impact of $5 billion from 2000 to 2010 within the Great Lakes Basin alone. The discovery of Asian carp near waterways connected to the lakes threatens to increase this problem.[2]

In response to these concerns, President Obama proposed $475 million in his fiscal year 2010 budget request for a new interagency initiative to accelerate the restoration of the Great Lakes. Specifically, the President requested that the U.S. Environmental Protection Agency (EPA) and its federal partners coordinate state, tribal, local, and industry actions to protect, maintain, and restore the integrity of the Great Lakes. The Department of the Interior Appropriations Act for fiscal year 2010 authorized EPA to transfer up to $475 million appropriated for the Great Lakes Initiative to any federal agency to carry out activities in support of the Great Lakes Restoration Initiative (GLRI).[3] The accompanying conference report directed EPA to develop a comprehensive, multiyear restoration action plan that includes targets and measurable objectives from fiscal years 2010 to 2014 that would lead to the restoration of the Great Lakes.[4] The conference report also directed EPA to provide detailed, yearly program accomplishments starting in 2011 and to develop a process that ensures monitoring and reporting on the progress of the GLRI. Nearly $300 million was appropriated in both fiscal years 2011 and 2012 for the GLRI and, according to EPA, about $284 million in fiscal year 2013, for a total of approximately $1.3 billion since fiscal year 2010.[5]

The Great Lakes Interagency Task Force (Task Force)—created in May 2004 by Executive Order 13340—oversees the GLRI.[6] The Task Force is chaired by the EPA Administrator, and it includes senior officials from the U.S. Departments of Agriculture, Commerce, Defense, Health and Human Services, Homeland Security, Housing and Urban Development, the Interior, State, and Transportation and the White House Council on Environmental Quality. The GLRI is implemented by EPA's Great Lakes National Program Office in conjunction with subagencies within the 11 Task Force agencies.[7] EPA and the other Task Force agencies issued the *Great Lakes Restoration Initiative Action Plan Fiscal Years 2010-2014* (Action Plan) in February 2010.[8] In addition, EPA created the Great Lakes Accountability System

(GLAS) as a mechanism for collecting information to monitor GLRI projects and progress.

Both our work and that of EPA's Office of the Inspector General have documented concerns about Great Lakes restoration efforts in the past. For example, we reported in May 2002 that the pace of cleaning up the Great Lakes and restoring areas of concern was slower than anticipated by the United States and Canada.[9] In September 2004, we found that EPA monitoring efforts did not provide comprehensive information on the condition of the Great Lakes and that monitoring by other federal and state agencies yielded information that that did not cover all areas related to the condition of the Great Lakes or the entire Great Lakes Basin.[10] In September 2009, EPA's Office of the Inspector General reported that EPA had not developed an effective management framework to clean up contaminated sediments in the Great Lakes areas of concern.[11]

In light of these past issues and the significant federal funds targeted toward addressing Great Lakes issues, you asked us to review the GLRI. This report examines (1) how the GLRI is implemented by the Task Force agencies and other stakeholders; (2) the methods that EPA has in place to assess GLRI progress; (3) the progress identified by the Task Force agencies and nonfederal stakeholders; and (4) the views of nonfederal stakeholders on factors, if any, that may affect or limit GLRI progress.

To examine how Task Force agencies and other stakeholders implement the GLRI and identify methods that EPA has in place to assess GLRI progress, we analyzed key documents including the Action Plan, EPA's GLRI financial reports, and information about GLRI projects from GLAS. In addition, we interviewed federal stakeholders—officials from each Task Force agency and relevant subagencies responsible for implementing the GLRI—and nonfederal stakeholders; specifically, representatives from state and local governments, tribes, nongovernmental organizations— such as the Nature Conservancy and the Great Lakes Indian Fish and Wildlife Commission—and academic institutions that that have received GLRI funds to gain an understanding of the GLRI and how it is implemented.[12] We also interviewed representatives from a prominent Great Lakes interest group, the Healing Our Waters – Great Lakes Coalition, that has not received GLRI funds. We visited GLRI projects, and attended the Eighth Annual Great Lakes Restoration Conference in Cleveland in September 2012. In addition, we interviewed a nonprobability sample of 21 individuals with expertise in Great Lakes' subject matter to obtain their views about the Action Plan and Great Lakes restoration. The 21 subject matter experts who participated in these interviews included officials from state agencies, members of academia, and nongovernmental organizations. We selected these experts in part through recommendations made by nonfederal attendees at the Great Lakes Restoration Conference. Because we used a nonprobability sample, the information obtained from these interviews is not generalizable to other individuals with Great Lakes-related expertise but provides illustrative information.

To examine the progress identified by the Task Force agencies and nonfederal stakeholders and the views of nonfederal stakeholders on factors, if any, that may affect or limit GLRI progress, we reviewed key documents detailing Great Lakes restoration efforts. We also administered a web-based survey to each of the 205 nonfederal stakeholders that, as of October 2012, had received GLRI funds from a Task Force agency and had a project identified in GLAS. We conducted this survey to identify examples of GLRI progress and obtain stakeholder views on factors that might limit progress. Of the 205 stakeholders contacted, 176 completed the survey, for an overall response rate of 86 percent. We also asked the 21 subject matter experts we interviewed to provide their views about factors that may limit GLRI progress. Appendix I presents a more detailed description of our objectives, scope, and methodology.

We conducted this performance audit from June 2012 to September 2013 in accordance with generally accepted government auditing standards. Those standards require that we plan and perform the audit to obtain sufficient, appropriate evidence to provide a reasonable basis for our findings and conclusions based on our audit objectives. We believe that the evidence obtained provides a reasonable basis for our findings and conclusions based on our audit objectives.

BACKGROUND

The Great Lakes Basin spans more than 750 miles from east to west and, as shown in figure 1, encompasses nearly all of Michigan, and parts of Illinois, Indiana, Minnesota, New York, Ohio, Pennsylvania, Wisconsin, and Ontario, Canada. The U.S. shoreline along the five Great Lakes— Erie, Huron, Michigan, Ontario, and Superior—is approximately 4,530 miles long, which is more than 2,000 miles longer than the U.S. coastline on the Atlantic Ocean.

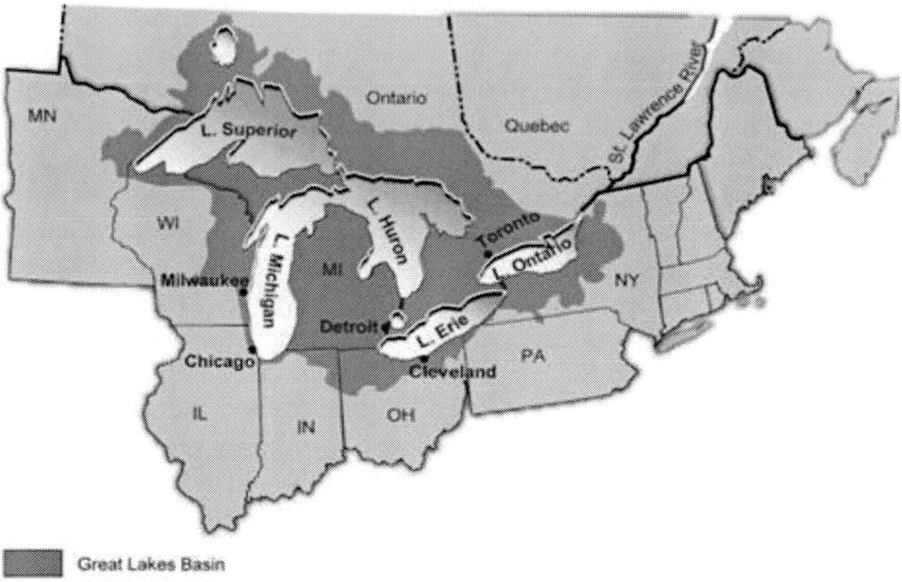

Source: GAO, Map Resources (map).

Figure 1. Area Comprising the Great Lakes Basin.

Numerous stresses threaten the health of the lakes themselves and the adjacent land within the Great Lakes Basin. For example, Asian carp threaten commercial and recreational fisheries in the Great Lakes because they tend to outcompete native fish for resources and modify habitat. In addition, despite progress in reducing the amount of phosphorus in the lakes achieved through mitigation techniques implemented in the 1970s, harmful algal blooms are once again threatening the Great Lakes Basin.[13]

The United States has taken several steps to restore the health of the Great Lakes ecosystem during the last four decades. Examples are as follows:

- In 1972, the United States and Canada signed the first binational Great Lakes Water Quality Agreement with the goal of restoring and maintaining the chemical, physical, and biological integrity of the Great Lakes Basin. The parties signed another Great Lakes Water Quality Agreement in 1978 that was amended in 1983, 1987, and in 2012. The 1987 amendment resulted in the formal identification of specific areas of concern, which were defined as geographic areas where a change in the chemical, physical, or biological integrity of the area is sufficient to cause restrictions on fish and wildlife or drinking water consumption, or the loss of fish and wildlife habitat, among other conditions, or impair the area's ability to support aquatic life.[14] The 2012 amendment contains provisions for addressing the nearshore environment, [15] aquatic invasive species, habitat degradation, and the effects of climate change, among other things.

- In 2002, the Great Lakes Legacy Act authorized EPA to, among other things, carry out sediment remediation projects and to conduct sediment contamination remediation research in areas of concern in the Great Lakes.[16]
- In 2004, the Task Force agencies collaborated with Great Lakes governors, mayors, tribes, and nongovernmental organizations in an effort referred to as the Great Lakes Regional Collaboration, which led to the development in 2005 of the *Great Lakes Regional Collaboration Strategy to Restore and Protect the Great Lakes*.[17] More than 1,500 individuals participated in this effort. A 2007 Brookings Institution report estimated that the $26 billion dollar investment necessary to implement this strategy would have resulted in $30 to $50 billion in short-term benefits to the regional economy and at least $50 billion in the long term.[18]

Even with these actions, the Great Lakes are environmentally vulnerable. In 2009, the President requested and Congress appropriated $475 million for fiscal year 2010 to create the GLRI. The accompanying conference report directed EPA to develop a GLRI Action Plan and to ensure that the GLRI funds supplement and expand, not supplant, federal agency Great Lakes programs.[19] EPA and the Task Force agencies issued the Action Plan in February 2010 to guide the implementation of the GLRI. According to the Action Plan, the GLRI is intended to build on previous restoration strategies to develop a collaborative approach to restoring the health of the Great Lakes ecosystem. The conference report also directed EPA to engage an independent, scientific panel to review the scientific credibility of the plan to optimize the likelihood of successful restoration at appropriate scales.[20] In response, EPA engaged its Science Advisory Board to conduct a review of the Action Plan.[21] Among other things, the Science Advisory Board recommended that EPA take steps to develop an adaptive management framework for the GLRI and noted that EPA should consider potential impacts of climate change on restoration.[22]

In March 2013, the Chair of the Council on Environmental Quality announced that the administration is committing to another 5-year GLRI Action Plan, for fiscal years 2015 to 2019. In May 2013, EPA and the Task

Force agencies began conducting public meetings and webinars to obtain comments on how the Action Plan could be refined to increase the effectiveness of GLRI investments during the next phase of the Action Plan.

TASK FORCE AGENCIES USE THE ACTION PLAN AND AGREEMENTS AMONG THEMSELVES AND WITH NONFEDERAL STAKEHOLDERS TO IMPLEMENT THE GLRI

The Task Force agencies use the Action Plan to guide implementation of the GLRI. They also use an interagency process to identify and transfer funds to GLRI work, and to enter into agreements among themselves and with nonfederal stakeholders to identify and implement GLRI projects, or they do the work themselves.

Task Force Agencies Use the Action Plan to Implement the GLRI

The Task Force agencies use the Action Plan to guide implementation of the GLRI. The Action Plan is organized into five focus areas that, according to the Task Force agencies, encompass the most significant environmental problems in the Great Lakes: (1) toxic substances and areas of concern; (2) invasive species; (3) nearshore health and nonpoint source pollution; (4) habitat and wildlife protection and restoration; and (5) accountability, education, monitoring, evaluation, communication, and partnerships. Table 1 describes each focus area.

Each focus area includes a description of its stresses and several longterm goals to address them. For example, one long-term goal in the invasive species focus area is to eliminate the introduction of new invasive species to the Great Lakes Basin. In addition, each focus area includes a number of objectives to be completed within the 5-year period covered by the Action Plan. For example, in the habitat and wildlife protection and restoration focus area, one objective is to assess 100 percent of U.S. coastal wetlands in the Great Lakes Basin by 2014.

The focus areas also include measures of progress—28 in total—each of which has annual targets for fiscal years 2010 to 2014 that are designed to ensure that efforts are on track to meet the long-term goals. Some of the measures address environmental conditions. For example, one measure for the habitat and wildlife protection and restoration focus area addresses the number of fish passage barriers that are to be removed or bypassed annually for the period of time covered by the Action Plan. The annual targets for the measure are the removal or bypassing of 100 barriers in 2010, 150 in 2011, 250 in 2012, 350 in 2013, and 450 barriers in 2014.

Table 1. Descriptions of the Five Great Lakes Restoration Initiative Action Plan Focus Areas

Focus area	Description
Toxic substances and areas of concern	Includes pollution prevention and cleanup of the most polluted areas in the Great Lakes
Invasive species	Includes efforts to institute a "zero tolerance policy" toward new invasions of non-native species, such as Asian carp
Nearshore health and nonpoint source pollution	Includes targeted geographic focus on high-priority watersheds and reducing polluted runoff from urban, suburban, and agricultural sources
Habitat and wildlife protection and restoration	Includes revitalizing wetlands and other habitat, and a comprehensive assessment of the entire Great Lakes coastal wetlands for the purpose of strategically targeting restoration and protection efforts in a science-based manner
Accountability, education, monitoring, evaluation, communication, and partnerships	Includes the implementation of goal- and results-based accountability measures, learning initiatives, outreach, and strategic partnerships

Source: GAO analysis of Great Lakes Restoration Initiative Action Plan.

The last part of each focus area is the principal actions, broad statements of the most significant activities that EPA and its federal partners conclude need to be done in order to achieve the goals, objectives, and targets in the Action Plan. For example, a principal action for the invasive species focus area is to promote the development and use of new control technologies, including biological control methods, that will significantly reduce the cost or increase the effectiveness of invasive species control measures.

See appendix III for a complete list of the Action Plan long-term goals, objectives, and measures of progress.

Task Force Agencies Use Interagency Process to Identify and Fund GLRI Work

The Task Force agencies identify the GLRI projects they plan to implement through an interagency process. EPA and the other Task Force agencies then enter into interagency agreements that describe the scope of the GLRI work that is to be undertaken and the amount of GLRI funds EPA will transfer to the Task Force agency doing that work.

Table 2. Number of Great Lakes Restoration Initiative Projects Funded by Task Force Agencies Reported in the Great Lakes Accountability System as of May 2013

Agency	Subagency	Number of projects
Environmental Protection Agency		498
Department of the Interior	Fish and Wildlife Service	379
	U.S. Geological Survey	75
	Bureau of Indian Affairs	33
	National Park Service	31
Department of Defense	U.S. Army Corps of Engineers	244
Department of Commerce	National Oceanic and Atmospheric Administration	113
Department of Agriculture	Forest Service	37
	Natural Resources Conservation Service	18
	Animal and Plant Health Inspection Service	18
Department of Homeland Security	U.S. Coast Guard	19
Department of Transportation	Federal Highway Administration	7
	Maritime Administration	4
Department of Health and Human Services	Agency for Toxic Substances and Disease Registry	5
Council on Environmental Quality		2
Total		1,483

Source: GAO analysis of EPA data.

The Task Force agencies and nonfederal stakeholders have started more than 1,450 GLRI projects since the program began.[23] GLRI projects can range in size from 0.25 acres for a fish spawning project to 10 million acres for a project to update habitat and wetland maps to identify sensitive and restorable wetlands across the basin. These projects may take place in one location or across multiple states within the Great Lakes Basin. GLRI projects have also ranged in cost from $3,000 to protect nesting piping plovers to $60.5 million to address contaminated sediment in multiple areas of concern.[24] Many GLRI projects take several years to complete and, as a result, many projects funded in fiscal years 2010 to 2012 are still under way. For example, approximately 64 percent of the fiscal year 2010 to 2012 projects reported in GLAS were at least half completed as of May 2013. Tables 2 and 3 show the number of projects funded by Task Force agency and by focus area as of May 2013, respectively.[25]

Table 3. Number of Great Lakes Restoration Initiative Projects by Focus Area Reported in the Great Lakes Accountability System as of May 2013

Focus area	Number of projects
Toxic substances and areas of concern	235
Invasive species	192
Nearshore health and nonpoint source pollution	247
Habitat and wildlife protection and restoration	630
Accountability, education, monitoring, evaluation, communication, and partnerships	183
Total	1,487

Source: GAO analysis of EPA data.

Although we did not break down GLRI funding per project in detail, some Task Force agency officials told us that they use a small percentage of that funding for administrative tasks. For example, officials from one Task Force agency told us that they used 5 percent of GLRI funds for one agency project for overhead and indirect costs.

In this case, the agency performs laboratory and monitoring work that the nonfederal stakeholders that implement the project do not have the capacity to do. As shown in figures 2 and 3, most GLRI funding has been utilized by EPA, the Department of the Interior's Fish and Wildlife Service, and the U.S. Army Corps of Engineers, and focus area one—toxic substances and areas of concern—has received more GLRI funding than the other focus areas in each of the past 4 years.

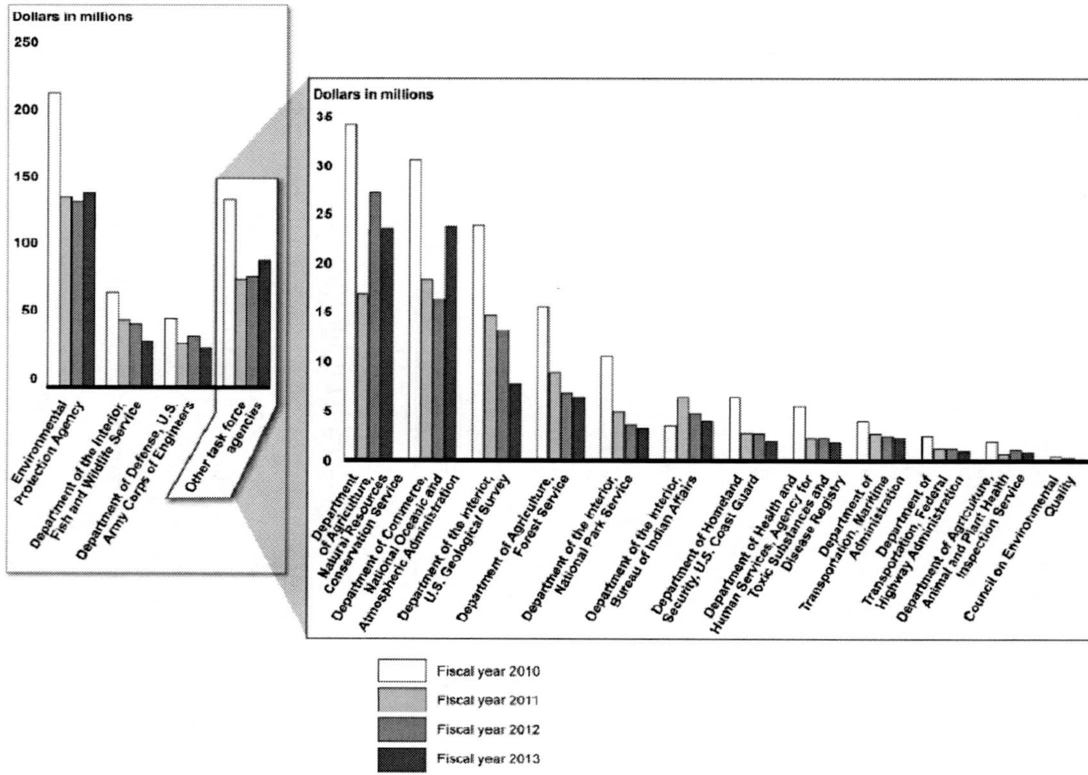

Source: GAO analysis of EPA data.

Note: In addition to funding Great Lakes Restoration Initiative (GLRI) projects, EPA uses its GLRI funds for, among other things, Great Lakes Legacy Act projects and GLRI funding for the Great Lakes Fishery Commission and International Joint Commission.

The Great Lakes Fishery Commission and International Joint Commission are binational efforts supported by the Department of State. The Great Lakes Fishery Commission works to sustain productivity of fish stocks of U.S. and Canadian concern in the Great Lakes, among other things, and the International Joint Commission assists the United States and Canada in the protection of the transboundary environment in part through the implementation of the Great Lakes Water Quality Agreement, among other things.

Figure 2. Great Lakes Restoration Initiative Allocations by Agency in Fiscal Years 2010 through 2013 as of April 2013.

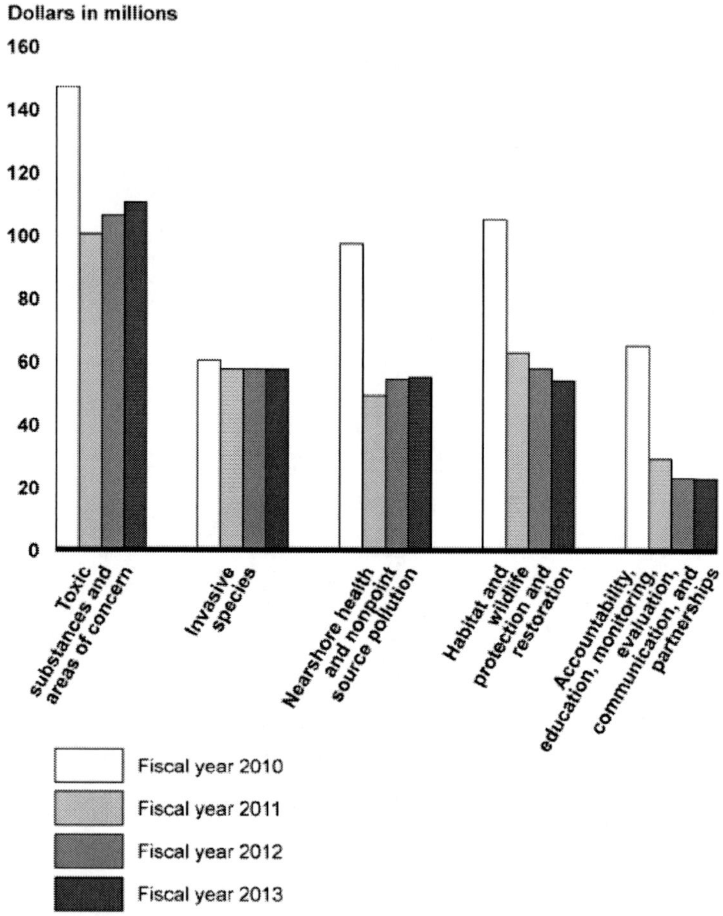

Source: GAO analysis of EPA data.

Figure 3. Great Lakes Restoration Initiative Allocations by Focus Area in Fiscal Years 2010 through 2013 as of April 2013.

EPA and the Task Force agencies have 2 years to obligate GLRI funds, which means that the amounts allocated in fiscal years 2010 and 2011 are not likely to change, but fiscal year 2012 and 2013 amounts may change as the period for obligating those funds comes to an end. In addition, while nearly all of the GLRI appropriations from fiscal years 2010 to 2012 have been obligated, not all of those obligations have been outlayed because many GLRI projects take several years to complete.[26] Table 4 shows the extent to which GLRI appropriations for fiscal years 2010 to 2012 have been obligated and outlayed as of July 2013.

Table 4. Percentage of Great Lakes Restoration Initiative Appropriations Allocated, Obligated, and Outlayed for Fiscal Years 2010 through 2012, as of July 2013

Dollar in millions			
Appropriations allocated, obligated, and outlayed	**Fiscal year 2010**	**Fiscal year 2011**	**Fiscal year 2012**
Appropriations	$475	$300	$300
Percentage of appropriations allocated	100%	99.8%	99.84%
Percentage of appropriations obligated	99.74%	99.74%	98.42%
Percentage of appropriations outlayed	79.53%	63.12%	28.17%

Source: GAO analysis of EPA information.

Note: Appropriation amounts are approximate and do not reflect across the board rescissions and other adjustments.

After EPA and the Task Force agencies have agreed to the GLRI fund transfers to each agency and the work that will be done, each agency either does the work itself or implements it through mechanisms such as cooperative agreements, contracts, or grants with nonfederal stakeholders. In the case of GLRI grants, a Task Force agency announces the availability of the grants through a request for application process. For example, EPA announced on April 19, 2012, that the agency planned to award approximately $20 million for about 100 projects in four of the five focus areas, and that the applications were due on May 24, 2012.[27] The request for application document included the amount of money and number of grants that would be awarded in each focus area and grant eligibility information, among other things. Similar announcements were made by other Task Force agencies for their GLRI grants, and EPA issued its first request for applications for fiscal year 2013 projects in July 2013.

EPA's Methods to Assess GLRI Progress May Not Produce Comprehensive and Useful Assessments or Support Program Adjustments

Although EPA uses information from GLAS and a variety of other sources to assess GLRI progress in meeting the annual targets associated with each of the 28 measures of progress in the Action Plan to assess GLRI progress, this information may not produce sufficiently comprehensive or useful assessments or support program adjustments. Specifically, (1) the Action Plan measures of progress currently in place may not provide sufficiently comprehensive or useful information and (2) EPA and the Task Force agencies have not fully established a plan to guide an adaptive management process for the GLRI that could allow them to assess the effectiveness of GLRI actions and, if needed, adjust their efforts.

EPA Uses the Action Plan Measures to Assess GLRI Progress

EPA assesses GLRI progress primarily by evaluating performance toward meeting the annual targets for the 28 measures of progress in the Action Plan, which EPA officials said are intended as indicators of success for each focus area. Of the 28 measures, 15 are also used in the agency's performance plan and performance report, which are required by the Government Performance and Results Act of 1993 (GPRA) as amended.[28] According to EPA officials, the agency sought to develop, in consultation with the Office of Management and Budget, a limited set of measures that could be used to evaluate GLRI progress on an annual basis. EPA officials said that, in developing the set of measures, they considered such factors as the availability of relevant data, the suitability of the data for measuring year-to-year changes, a focus on environmental outcomes, and the existing federal authorities available to address the measure. Officials from EPA and another Task Force agency also told us that many of the measures were based on existing baseline information and were practical because they used existing agency databases for information.

To gauge GLRI progress toward meeting the targets for the measures, EPA obtains the data needed using information from GLAS, EPA programs, other federal agencies, states, and universities. For example, for 7 of the 28 measures of progress, GLAS provides progress data specific to individual GLRI projects. In addition, states provide information on progress at areas of concern, the National Oceanic and Atmospheric Administration provides progress information from a database it manages on invasive species in the Great Lakes, and the Fish and Wildlife Service provides progress information related to habitat and wildlife protection and restoration. In March 2013, the Task Force agencies issued the first GLRI progress report to Congress, which identified fiscal year 2011 progress using the measures for each focus area and described the accomplishments of some projects.[29] The Task Force agencies also issued a report on fiscal year 2010 progress in April 2013.[30]

Action Plan Measures of Progress May Not Provide Sufficiently Comprehensive or Useful Information

Assessments of GLRI progress could help GLRI stakeholders, Congress, and the public discern the extent to which the health of the Great Lakes ecosystem has been restored, as well as what has been achieved by the approximately $1.3 billion appropriated to the program since fiscal year 2010. However, the Action Plan measures of progress may not provide sufficiently comprehensive or useful information on GLRI progress because (1) some goals and objectives do not link to any measures that will allow EPA to assess progress toward achieving them; (2) the measures do not capture the results of many of the GLRI projects; (3) data used to evaluate some measures of progress may not be complete; and (4) some Task Force and state agency officials, subject matter experts, and EPA's Science Advisory Board raised concerns about the usefulness of some measures or their targets for indicating progress toward the GLRI goals and objectives.

Some GLRI Goals and Objectives Do Not Link to Any Measures

The Action Plan does not identify the links between a focus area's goals, objectives, and measures. Specifically, based on our analysis of the GLRI Action Plan, some of the goals and objectives in the Action Plan do not link to any measures of progress, which EPA uses to assess GLRI progress. For example, one of the goals of the habitat and wildlife protection and restoration focus area is that development activities are planned and implemented in ways that are sensitive to environmental considerations and compatible with fish and wildlife and their habitats. However, we found that none of the six objectives and nine measures of progress in this focus area link to this goal. Similarly, in its 2012 review of the GLRI Action Plan, EPA's Science Advisory Board commented on the ambiguity in the links between the elements of the plan, such as between the long-term goals and the objectives. The Science Advisory Board also recommended that the Action Plan tie the measures more directly to the goals, noting that the more clear and transparent the connection between the measures and the goals, the easier it will be to document how well the actions are working to meet the goals.

EPA officials told us that the Action Plan was not intended to contain direct linkages between every goal, objective, and measure but rather that the measures were to serve as key indicators of success for each focus area. They said that the rationale for that decision was to provide some flexibility, recognizing that not all important restoration work may fall within the measures of progress and that the Action Plan does not represent the entirety of actions necessary for Great Lakes protection and restoration. However, based on our review of documents on planning for restoration of natural resources, we believe that clear linkages between a plan's goals, objectives, and measures are critical to achieving and assessing progress over time. For example, the National Research Council recommended such linkages in its checklist for planning and evaluating aquatic ecosystem restoration.[31] Specifically, the National Research Council recommended constructing specific performance indicators that are directly and appropriately linked to each objective, noting that these performance indicators are specific, measureable quantities that reveal to what extent the objectives are being achieved.[32] In addition, the Department of the Interior's 2009 technical guidance on managing natural resources emphasizes that, to be useful for decision making and evaluation, objectives need to be unambiguous, with specific metrics and specific target conditions.[33]

We recognize that it may not always be feasible or appropriate to identify measures for every aspect of each goal or objective. Nonetheless, although some objectives in the Action Plan describe quantifiable actions, the Action Plan contains no measures of progress for them. For example, two Action Plan objectives for the nearshore health and nonpoint source pollution focus area state that, by 2014, (1) 50 percent of high priority Great Lakes beaches will have been assessed using a standardized sanitary survey tool to identify sources of contamination and (2) rapid testing or predictive modeling methods (i.e., to improve the accuracy of decisions on beach postings to better protect public health) will be employed at 33 percent of high-priority beaches. However, because the Action Plan contains no measures of progress that link to these objectives, it is unclear how EPA will be able to assess annual progress toward these and other objectives that do not have linked measures. EPA officials told us that they recognize the need to report a more comprehensive assessment of objectives and are considering options for evaluating and reporting on them.

Measures Do Not Capture the Results of Many GLRI Projects

Many of the projects funded by the GLRI do not have an Action Plan measure of progress assigned to them, which means that the results of those projects are not captured by the measures. Specifically, nearly 60 percent of the more than 1,450 GLRI projects reported in GLAS as of May 2013 did not have an associated Action Plan measure of progress. One reason for this is that, according to a Task Force agency, some projects contribute to the Action

Plan measures but are not the type of on-theground restoration projects that are addressed by most of the measures of progress. Officials from the U.S. Geological Survey told us that this is the case for much of their agency's GLRI work. For example, the agency's work has included efforts to develop new methods of controlling phragmites, an invasive plant.[34] This work contributes toward the Action Plan's objective of developing or refining and pilot testing five technologies to contain or control invasive species by 2014. However, there is no linked measure of progress for this objective or assigned to the project. As a result, EPA's progress assessments may inaccurately capture the extent of progress being made toward containing or controlling invasive species.

According to EPA officials, projects without assigned measures address the Action Plan goals, objectives, or principal actions; the reason these projects do not have measures assigned to them is that they do not directly provide data for any of the Action Plan's 28 measures. They told us that they monitor the results of individual projects through standard agency practices, but that they are not currently tracking cumulative results that are not addressed by an Action Plan measure of progress. According to agency officials, future GLRI progress assessments may capture more information. Specifically, they said that the cumulative results of some of the projects without assigned measures of progress may ultimately be captured by some of the existing measures. We recognize that this may be the case for some projects. For example, projects to restore native fish habitat that do not have assigned measures of progress could help lead to progress over time toward the measure that addresses the percentage of native aquatic species populations that are self-sustaining in the wild. However, a cumulative approach may not allow EPA to capture specific progress information from those projects without measures, and EPA uses the measures to assess GLRI progress and has been directed to report on that progress annually. As we reported in November 2002, one attribute of successful performance assessment is that there should be enough measures to ensure that an agency has the information it needs about project performance.[35] Without methods to include the results of such projects, EPA cannot ensure that the agency is assessing the full extent of progress being made.

Data Used to Evaluate Some Measures of Progress May Not Be Complete

Data used to evaluate some measures of progress may not be complete because GLAS users are limited to reporting progress using a single measure, and GLRI projects may directly address multiple measures across different focus areas. We found that 7 of the 28 measures of progress are tracked primarily in GLAS and, as we noted previously, EPA uses information from GLAS, as well as other sources, to gauge GLRI progress toward meeting the targets for the measures. Officials from five Task Force agencies told us that this is a significant limitation that can result in underreporting of progress. For example, a National Park Service GLRI project has involved managing acres for invasive species, which is one measure, as well as outreach to the public on practices that prevent the introduction and spread of invasive species, which is another measure. However, GLAS requires the agency to choose only one of these measures for reporting progress. EPA officials told us that GLAS users are restricted to reporting on a single measure due primarily to a decision by EPA to simplify the reporting process during the initial stages of the GLRI. Agency officials said that this decision was made for several reasons, including concerns about double-counting and overreporting, a desire to minimize the reporting burden for GLAS users and EPA, and the need to ensure appropriate data quality review before making the progress information public. They said that the current design of GLAS reflects this decision and that system modifications would be necessary to allow reporting on multiple measures. However, by limiting users to reporting information on only a single measure for each project, GLAS prevents EPA from collecting complete progress information on GLRI projects—that is, information on each of the measures addressed by GLRI projects. As we noted previously, there should be enough measures to ensure that an agency has the information it needs to assess project performance. Without collecting information about the multiple measures affected by a project, the data EPA is using for certain measures of progress cannot be complete, and EPA is likely underreporting progress for these measures.

EPA officials told us that they began to address this issue in fiscal year 2012 for two complementary measures in the habitat and wildlife protection and restoration focus area, the miles of rivers reopened for fish passage, and the number of fish passage barriers removed or bypassed. Specifically, EPA has begun asking GLRI funding recipients reporting on one of these measures to indicate progress, if any, on the complementary measure. This effort will help EPA obtain more complete information on these two measures, but it does not address the broader reporting limitation in GLAS, which EPA officials told us may result in underreporting of progress for certain measures. EPA officials told us that they will consider addressing this limitation in GLAS, but they did not indicate a time frame for doing so.

In addition, although EPA officials told us that they have concerns about the quality of GLRI progress information in GLAS, they have not fully assessed the quality of that information, such as its completeness, accuracy, and consistency. As a result, the agency has not made this information available to the public. While practices required under GPRA, as amended, are required at the federal department or agency level, we have previously reported that these requirements can serve as leading practices for planning at lower levels within federal agencies, such as individual programs or initiatives.[36] Thus, EPA is not required to address requirements under GPRA, as amended, for GLAS, but by following them, the agency would be implementing leading practices. The GPRA Modernization Act of 2010 requires, among other things, that agency performance plans and reports describe how the agency ensures the reliability of the data used to measure progress toward its performance goals, including how it verifies and validates measured values of performance and compensates for any limitations to the data to reach the required level of accuracy.[37] Verification includes the assessment of data completeness, accuracy, and consistency to ensure that the data will be of sufficient quality to document performance and support decision making.[38] In light of this and EPA's concerns about the quality of the GLRI progress data in GLAS, we also have concerns about the use of these data for measuring progress toward the Action Plan goals and objectives.

GLRI funding recipients are responsible for entering information about the progress of their GLRI projects directly into GLAS, and EPA has methods in place to review this information on a project-by-project basis. For example, EPA has developed a plan for managing the grants, interagency agreements, and contracts the agency awards for GLRI projects. According to agency officials, this plan synthesizes existing agency standards and policies and itemizes the activities to be undertaken by project officers and others to ensure an effective grant oversight process. Among other things, this plan specifies that, for interagency agreements, EPA officials who are GLRI project officers will review progress reports submitted by GLRI funding recipients and that the project officers will compare the information in those reports to the information entered in GLAS. [39] According to EPA officials, project officers for grants are also expected to compare information in progress reports with information in GLAS.

However, EPA officials noted that they believe there are overall consistency issues with the progress information grantees enter in GLAS, and that there could be underreporting of progress for certain measures when they are being achieved by a large number of projects or stakeholders. However, the officials told us that they have not yet taken steps to identify the progress information in GLAS that may be incomplete, incorrect, or inconsistently entered by GLRI funding recipients. As a result, EPA's ability to reliably assess progress toward the targets for the seven measures tracked primarily in GLAS is questionable. EPA officials told us that they recognize the need to improve GLAS and that they are beginning efforts to review the system, including assessing their data quality and review procedures and identifying improvements. The officials noted that they are also considering whether GLAS is the right tool for the GLRI. EPA officials had not determined a time frame for this work as of June 2013.

Some Measures or Their Targets May Not Be Indicative of Progress

Some Task Force and state agency officials, subject matter experts, and EPA's Science Advisory Board, raised concerns that certain measures of progress or their targets may not be useful for indicating GLRI progress toward the Action Plan's goals and objectives.

Climatic Factors May Affect the Usefulness of Some Measures

Five of the six measures of progress for the nearshore health and nonpoint source pollution focus area may not be useful for indicating GLRI progress over the short term toward the Action Plan's goals and objectives. Specifically, EPA officials and others have noted that these measures address environmental conditions that are influenced by climatic factors, such as precipitation and wind patterns. These factors make it difficult to identify whether the restoration efforts are having the desired effects. For example, one of the measures of progress in the Action Plan for this focus area tracks the square miles of harmful algal blooms in the Great Lakes. According to information from EPA officials, such algal blooms are influenced by climatic factors such as precipitation patterns, water temperature, and wind speed and direction. Officials from another Task Force agency also told us that because of factors such as weather that cannot be controlled, the extent of algal blooms in any given year is not directly related to the management actions being taken in the GLRI and, as a result, the measure is not useful for indicating GLRI progress toward the Action Plan's goals and objectives. EPA officials told us that, given the short period of time the GLRI has been under way, factors such as temperature and the amount and timing of precipitation are currently the primary factors affecting the extent of algal bloom. Officials also told us that, over the long term, management actions will lead to lower phosphorous levels, which will have a minimizing affect on the extent of such algal blooms.

However, over the short term, this measure may not be useful for indicating GLRI progress toward the focus area's objective of significantly reducing the number and severity of incidences of harmful algal blooms by 2014. In addition, we have previously reported criteria for determining the extent to which an agency's performance plan provides a clear picture of intended performance, including that measures must clearly represent or be related to the performance they are trying to assess.[40] Therefore, measures that track conditions that are not directly related to management actions being taken may not be useful for indicating GLRI progress toward the Action Plan's goals and objectives. While we recognize that the Action Plan is not a performance plan, using criteria intended for performance plans is appropriate because EPA uses the Action Plan's measures of progress to assess GLRI progress as an agency would use the measures in a performance plan.

Some Measures Track Actions That May Not Be Sufficient to Lead to the Desired GLRI Goals

For example, one of the goals for the toxic substances and areas of concern focus area is that environmental levels of toxic chemicals are reduced to the point that all restrictions on the consumption of Great Lakes fish can be lifted. A measure of progress that links to this goal tracks the long-term reduction in average concentrations of PCBs in Great Lakes fish. However, officials from two Task Force agencies and two state agencies, as well as five subject matter experts, reported that the measure's focus on PCBs is too narrow and that other contaminants, particularly mercury, need to be addressed as well. EPA's Science Advisory Board also noted this narrow focus in its 2012 review of the Action Plan. Mercury is important because, according to a 2011 binational study of mercury in the Great Lakes region, it has widely contaminated the region and has been responsible for fish consumption advisories in the eight Great Lakes states and the Canadian province of Ontario. [41] The study concluded that mercury, largely due to atmospheric emissions from coal-fired power plants, remains a pollutant of major concern with an impact on fish in the region that is much greater than previously recognized. Consequently, reducing average concentrations of PCBs in fish is not likely to lead to lifting all restrictions on the consumption of Great Lakes fish. Because this measure does not clearly represent all of the contaminants that need to be addressed, it may not be useful for indicating GLRI progress toward the Action Plan's goals and objectives.

Monitoring Practices May Affect the Usefulness of Some Measures

For example, one goal of the invasive species focus area is to eliminate the introduction of new invasive species, with a 2014 objective of reducing the yearly average rate of invasive species newly detected in the Great Lakes ecosystem by 40 percent, compared with the period from 2000 to 2009. The linked measure of progress addresses the rate at which nonnative species are newly detected in the Great Lakes ecosystem. The source of data used to evaluate this measure is a database maintained by the National Oceanic and Atmospheric Administration. According to information on this database, the identification of new species depends on the ability to find, recognize, verify, and document new species, which is, in turn, dependent on the ability to adequately sample the Great Lakes ecosystem. Officials from a state agency and four subject matter experts raised concerns about the usefulness of this measure for assessing progress toward the Action Plan objectives, noting that the number of new species detected will increase if surveillance for invasive species increases or improves. One of these experts noted that meeting the targets may not represent progress because monitoring efforts are low, and another of these experts told us that the measure needs to be combined with a known level of monitoring. Similarly, National Oceanic and Atmospheric Administration officials agreed that progress toward the targets for this measure will vary depending on the extent of monitoring and the ability of surveillance to detect new nonnative species, and they told us that, without a known level of monitoring, it may not be possible to reliably identify trends in the introduction or detection of new species using this measure. As we reported in November 2002, one attribute of successful performance measures is that they are likely to produce the same results if applied repeatedly to the same situation.[42] If efforts to identify new species depend on the ability to adequately sample the Great Lakes ecosystem and surveillance or levels of monitoring are not consistent throughout the program, then this measure may not be useful for indicating GLRI progress toward the Action Plan's goals and objectives.

Targets for Some Measures May Not Represent Significant Ecological Improvement

EPA's Science Advisory Board reported in its 2012 review of the GLRI Action Plan that it is not clear whether the targets for the measures of progress reflect significant or measurable improvement, or whether achieving the targets will result in real ecological benefit. It also noted that it is not clear how the targets were developed and, that while the measures of progress include baselines for the targets, the universe is not always defined.[43] For example, one measure in the nearshore health and nonpoint source pollution focus area addresses the amount of land with certain conservation practices implemented to reduce erosion, nutrients, and pesticides. The Action Plan reports a baseline of 165,000 acres with such practices already being implemented and identifies a 2014 target of 247,500 acres, a 50 percent increase. However, the Action Plan does not identify the universe (i.e., the total acreage of land upon which such practices could be implemented), which the Science Advisory Board reported is more than 38 million acres of agricultural land in the United States within the Great Lakes Basin. Consequently, using this universe, the 50 percent increase in acreage using such conservation measures represents a change from 0.4 percent to 0.6 percent of the total U.S. agricultural land in the Great Lakes Basin. According to the Science Advisory Board, it is not clear if this change is meaningful and how this percentage of improvement will potentially result in the restoration of the Great Lakes ecosystem. In November 2002, we reported that clarity is a key attribute of successful performance measure.[44] Without some clarification, this measure may not be useful for indicating GLRI progress toward the Action Plan's goals and objectives.

A subgroup of the Task Force agencies has been evaluating how well the measures of progress and targets are working and has identified some that will need revisions. For example, the subgroup found that some measures of progress have been difficult to implement or difficult to demonstrate scientifically, particularly in the nearshore health and nonpoint source pollution focus area. However, as this subgroup has acknowledged, there is no defined process for the Task Force agencies to revise the Action Plan, such as updating or replacing the measures of progress or for updating their targets.

EPA and the Task Force Agencies Have Not Fully Established an Adaptive Management Plan for the GLRI

EPA and the Task Force agencies have not fully established a plan to guide an adaptive management process for the GLRI. Although there is no requirement that the Task Force establish an adaptive management plan for the GLRI, an adaptive management process could allow EPA and the Task Force agencies to evaluate whether GLRI projects are leading to the Action Plan's objectives and goals and, if needed, use the results to adjust future actions.

EPA and several Task Force agencies in 2000 adopted a unified federal policy on watershed management that defined adaptive management as a type of natural resource management in which decisions are made as part of an ongoing science-based process that involves (1) testing, monitoring, and evaluating applied strategies; (2) incorporating new knowledge into management approaches that are based on scientific findings and the needs of society; and (3) using results to modify management policy, strategies, and practices.[45]

This policy stated that the agencies would incorporate adaptive management principles into their programs and use adaptive management to improve watershed conditions. More recently, in the 2012 amendment to the Great Lakes Water Quality Agreement, the United States and Canada agreed to be guided by the principles and approaches of adaptive management, which is described in the agreement as a systematic process to assess the effectiveness of actions and adjust future actions to achieve the objectives of the agreement, as outcomes and ecosystem processes become better understood.

There is no universal definition for adaptive management or fixed set of steps that constitutes an adaptive management process, but key elements of this iterative process based on guidance documents from several Task Force agencies are summarized in figure 4. In addition, according to guidance from the Forest Service, adaptive management requires explicit designs that, among other things, specify documentation and monitoring protocols; roles, relationships, and responsibilities; and, assessment and evaluation processes. This guidance also notes that it is important for an adaptive management effort to have clear documentation describing details of the adaptive management process, and an absence of explicit plans can diminish the potential benefits of adaptive management.[46]

In its 2012 review of the Action Plan, EPA's Science Advisory Board found that the GLRI Action Plan did not establish an adaptive management framework and recommended that EPA develop a science-based plan that, when coupled with the Action Plan, would create an adaptive management framework for the GLRI. In response to this recommendation, in May 2013, the Task Force agencies released the draft *GLRI Adaptive Science-Based Framework for Great Lakes Restoration*, or draft framework, for public comment.[47]

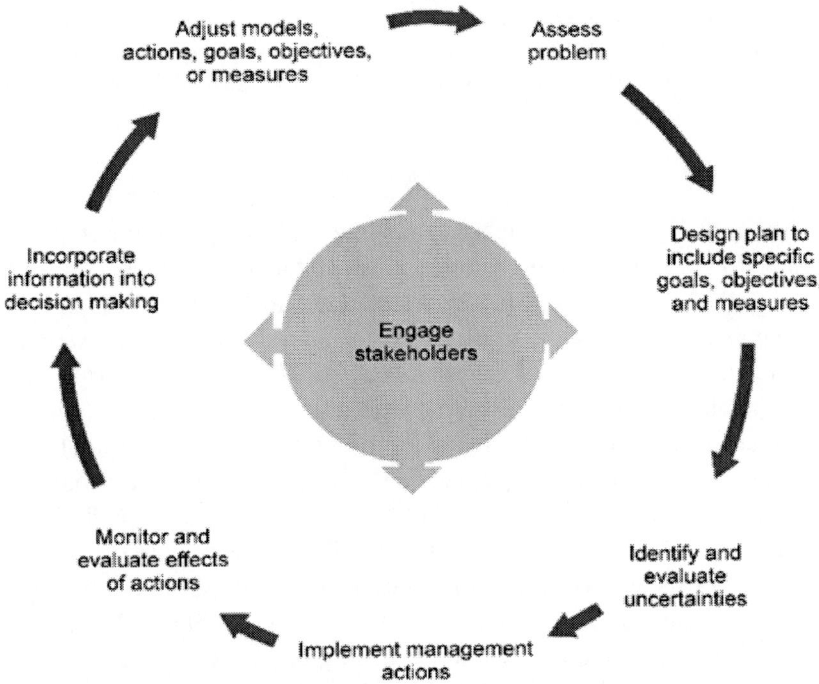

Sources: GAO analysis of guidance from the Department of the Interior,[a] U.S. Geological Survey.[b] U.S. Army Corps of Engineers.[c] Forest Service,[d] and EPA.[e]

[a]B. K. Williams and E. D. Brown, *Adaptive Management: The U.S. Department of the Interior Applications Guide*, Adaptive Management Working Group, U.S. Department of the Interior (Washington, D.C.: 2012).

B. K. Williams, R. C. Szaro, and C. D. Shapiro, *Adaptive Management: The U.S. Department of the Interior Technical Guide*, Adaptive Management Working Group, U.S. Department of the Interior (Washington, D.C.: 2009).

[b]A. J. Atkinson, P. C. Trenham, R. N. Fisher, S. A. Hathaway, B. S. Johnson, S. G. Torres, and Y. C. Moore, *Designing Monitoring Programs In an Adaptive Management Context For Regional Multiple Species Conservation Plans*, U.S. Geological Survey Technical Report (U.S. Geological Survey Western Ecological Research Center, Sacramento, CA: 2004).

[c]RECOVER 2010, *Comprehensive Everglades Restoration Plan Adaptive Management Integration Guide*, Restoration Coordination and Verification (Jacksonville, FL and West Palm Beach, FL: U.S. Army Corps of Engineers, Jacksonville District and South Florida Water Management District, 2010).

[d]G. H. Stankey, R. N. Clark, and B. T. Bormann, *Adaptive Management of Natural Resources: Theory, Concepts, and Management Institutions*, Gen. Tech. Rep. PNW-GTR-654 (Portland, OR: U.S. Department of Agriculture, Forest Service, Pacific Northwest Research Station, 2005).

[e]EPA, *Watershed Analysis and Management (WAM) Guide for States and Communities: EPA Watershed Analysis and Management Project* (Washington, D.C.: 2003), and EPA, Handbook for Developing Watershed Plans to Restore and Protect Our Waters, EPA 841-B-08-002 (Washington, D.C.: 2008).

Figure 4. Key Elements of the Adaptive Management Process.

The draft framework recommends use of an "adaptive restoration" approach for the GLRI that includes most of the key elements of adaptive management. According to the draft framework, adaptive restoration involves exploring alternative ways to meet the goals and objectives in the Action Plan, predicting the outcomes of alternatives based on the current state of knowledge, implementing one or more of these alternatives, monitoring to learn about the impacts of restoration actions, and incorporating new knowledge into restoration strategies that are based on scientific findings and the needs of society. However, the draft framework does not provide details on how the elements of adaptive management will be implemented. For example, neither the draft framework, nor the Action Plan, addresses how the agencies will complete the adaptive management loop by incorporating the information gained in future decision making and adjusting the GLRI, if needed. The draft framework indicates that re-assessment of the Action Plan goals, objectives, and measures of progress should be considered every 5 years. However, neither the framework nor the Action Plan outlines a process for assessing or incorporating new information or altering the Action Plan, such as by refining the measures or their targets, as recommended by the adaptive management section of EPA's 2003 guidance for states and communities on watershed analysis and management.[48] For example, neither the draft framework nor the Action Plan identifies decision thresholds or criteria for making changes, which the Science Advisory Board's 2012 report and EPA's 2003 guidance have identified as important. In addition, the draft framework does not include two of the key elements of adaptive management shown in figure 4: engaging stakeholders, and identifying and evaluating uncertainties.

Engaging Stakeholders

According to guidance from several Task Force agencies, active and sustained stakeholder involvement is essential for effective implementation of adaptive management.[49] The draft framework does not explicitly include stakeholder engagement as an element of the adaptive restoration process. It acknowledges that engaging scientists, stakeholders, and the general public should be considered for updating the GLRI over time and states that interactive engagement in setting goals and objectives is a key element for producing results. Nevertheless, it does not address how the Task Force agencies will ensure ongoing stakeholder engagement in the GLRI. Without explicit plans in the draft framework for engaging stakeholders, it is not clear how the Task Force agencies will consider and account for potential changes in stakeholder perspectives and priorities. As the Department of the Interior's guidance notes, conflicting priorities among stakeholders can be enough to prevent the successful implementation of adaptive management.[50] This guidance also explains that differing stakeholder views about how natural processes work and how they respond to management are examples of uncertainties that can limit the effectiveness of management efforts.

Identifying and Evaluating Uncertainties

Neither the draft framework, nor the Action Plan, identifies, prioritizes, and evaluates critical scientific or policy uncertainties. In addition, the draft framework does not include such uncertainties in its list of elements of adaptive restoration planning. Such uncertainties could include critical assumptions, gaps in knowledge, or uncertainties about the relationships between ecological processes and stresses, such as the potential effects of climate change on the Great Lakes. Identifying and explicitly accounting for critical uncertainties in designing the GLRI could increase the likelihood that the GLRI will meet its goals and objectives. Several Task Force agencies have emphasized the use of models as an important way to evaluate and account for such uncertainties by, for example, representing how the ecosystem may respond to restoration actions or environmental changes.[51] The draft framework notes that predictive modeling plays an important role in adaptive restoration, and it recommends evaluation of uncertainties in the context of reporting on progress.[52] However, the framework does not identify and account for critical uncertainties.

TASK FORCE AGENCIES AND NONFEDERAL STAKEHOLDERS REPORT GLRI PROGRESS, BUT OVERALL GREAT LAKES RESTORATION PROGRESS IS DIFFICULT TO QUANTIFY

The Task Force agencies have issued two GLRI progress reports, but the reports include few specific examples of progress. To obtain further insights about GLRI progress, we surveyed nonfederal GLRI stakeholders and interviewed Task Force agency officials. However, quantifying overall Great Lakes restoration progress is difficult for several reasons, including the unique environmental conditions of each of the Great Lakes.

Task Force Agencies Report Progress

As we noted previously, the Task Force agencies have issued two GLRI progress reports to Congress; one on fiscal year 2011 progress and one on progress in fiscal year 2010. EPA officials told us that the report on progress in fiscal year 2011 includes more information and was released earlier than the fiscal year 2010 report because GLRI funding was available late in fiscal year 2010 and not as much was achieved that year. Both reports identified whether the targets for the GLRI Action Plan's 28 measures of progress had been met or exceeded in the related fiscal year, or that data were unavailable for a specific measure. For example, the fiscal year 2011 report states that 15 measures were met or exceeded, 9 measures were not met, and data were not available to determine the status of 4 measures. However, as we noted earlier, the measures may not provide sufficiently comprehensive or useful information for a number of reasons, including that the measures do not capture the results of many of the GLRI projects and that some measures or their targets may not be indicative of progress.

The fiscal year 2011 report also includes broad statements about whether the GLRI is achieving goals, objectives, and measures of progress that are followed by several examples of results and highlights of progress made or expected in two projects in each of the five Action Plan focus areas.[53] For example, in the invasive species focus area, the report states that:

> In the first years of the GLRI, no new aquatic invasive species populations have been detected in the Great Lakes. The GLRI is at the forefront of invasive species prevention, control, and rapid response. The GLRI is supporting investments in technologies that prevent the introduction of invasive species, including the Department of Transportation's Maritime Administration's verification of new ballast water treatment technologies, which is an important step before conducting ship-scale testing.

However, except for the two projects highlighted in each focus area, the fiscal year 2011 report does not identify the GLRI projects that led to the progress achieved. In addition, neither of the two reports integrates accomplishment and progress information into conclusions about overall GLRI progress or contains specific information about the extent of progress toward GLRI goals, and neither includes an assessment of changes in the health of the Great Lakes ecosystem that could be attributed to the projects. As a result, the information in the GLRI progress reports was not sufficient for determining if GLRI projects had led to the described progress and how much progress has been made in restoring the health of the Great Lakes ecosystem.

Nonfederal Stakeholder and Task Force Agency Examples of GLRI Progress

In light of the limited information included in the two GLRI progress reports, we sought to obtain further insights by surveying nonfederal stakeholders and interviewing officials from Task Force agencies. The information we gathered in the survey allows us to provide more information on how nonfederal stakeholders describe GLRI progress, and it is not intended to determine the extent of GLRI progress toward restoring the health of the Great Lakes ecosystem. When asked to provide examples of how one or more of their organization's GLRI projects had benefitted the Great Lakes ecosystem, 87 percent (153) of the 176 respondents provided at least one example of how one or more of their GLRI projects had benefited, or was expected to benefit, the Great Lakes ecosystem in each of the five focus areas. Specifically:

- 55 percent (96) of respondents provided at least one example of how one or more of their GLRI projects had directly benefited the Great Lakes ecosystem, such as by restoring habitat or by trapping invasive species.
- 27 percent (47) of respondents provided at least one example of how one or more of their GLRI projects indirectly benefited the Great Lakes ecosystem, such as by improving the understanding of part of the ecosystem or identifying future restoration work.[54]
- 44 percent (77) of respondents reported that they expected at least one of their projects to result in a direct or indirect benefit to the ecosystem, but that it was too early to report a benefit at this time.

- 13 percent (23) of respondents did not provide an example of how one of their projects had, or was expected to, benefit the Great Lakes ecosystem.

A majority of the survey respondents and Task Force agency officials we interviewed reported progress toward the overall GLRI goal of restoring the health of the Great Lakes ecosystem. The following examples represent a broad range of GLRI progress in each of the five focus areas as reported by survey respondents, through the survey and follow-up interviews, and by Task Force agency officials, in interviews, and do not represent the only types of GLRI projects being implemented:

Toxic Substances and Areas of Concern

- One survey respondent reported that her organization received two GLRI grants to help clean up the Presque Isle Bay area of concern in Lake Erie, in Pennsylvania; the two grants totaled $664,789. The respondent told us that, prior to the GLRI, the organization was devoting all of its resources to addressing only one of the specific conditions—fish tumors—that were responsible for the bay being identified as an area of concern, and that GLRI funding made it possible for her organization to do additional work. Specifically, GLRI funding enabled the organization to undertake a number of monitoring and investigative projects; develop a target for being removed from the list of areas of concern; and, develop a long-term monitoring, restoration, and protection plan for the bay's watershed. The respondent reported that GLRI funding was also used for specific investigations in the watershed that will provide her organization with an updated set of data to use in identifying priority areas for restoration and protection. According to the respondent, without that information, the organization would not have a blueprint to maintain and continue protecting the bay. In December 2012, 21 years after the bay had been identified as an area of concern, the survey respondent told us in an interview that her organization was in the process of sending a formal request to remove Presque Isle Bay's designation as an area of concern. EPA announced in February 2013 that the bay had been delisted. It is the second U.S. area of concern to be removed from the list. The respondent estimated that GLRI funding had accelerated the process of removing Presque Isle Bay from the list of areas of concern by 10 years.

- Officials we interviewed from the U.S. Geological Survey told us that the agency had used GLRI funds to complete three basin-wide sampling efforts of the water column, bottom sediments, and two other types of samples across the entire Great Lakes Basin as a part of its methylmercury sampling and analysis project. Methylmercury—an organic form of mercury—is a highly toxic substance that can build up in predatory fish, including fish that people eat. According to the officials, the sampling efforts took place in August 2010, April 2011, and August 2011, and data from these efforts have revealed a previously unknown source of methylmercury that likely is the dominant source leading to elevated concentrations in fish throughout the Great Lakes.

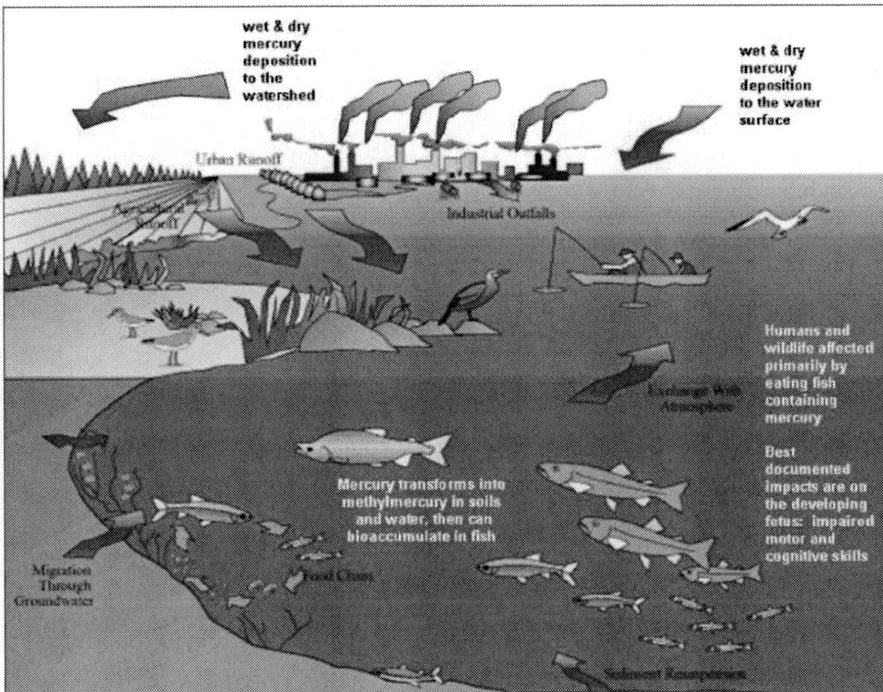

Source: National Oceanic and Atmospheric Administration.

Figure 5. A Depiction of the Mercury Cycle.

Mercury

Mercury is a highly toxic metal that can pose health risks to wildlife and people. Specifically, high levels of mercury exposure can harm the brain, heart, kidneys, lungs, and immune system of people of all ages. People are most commonly exposed to an organic form of mercury called methylmercury. Methylmercury accumulates up the food chain, so big fish that eat smaller fish tend to contain more methylmercury. People are primarily exposed to methylmercury through eating contaminated fish. According to a 2011 Biodiversity Research Institute report, mercury pollution is ubiquitous across the Great Lakes region, primarily as a result of emissions from coal-fired power plants, and five states in the region have issued statewide consumption advisories for mercury in fish from all freshwaters, two have issued statewide advisories for mercury in fish from all lakes, and one has issued advisories for specific water bodies.

Notes: D.C. Evers, J.G. Wiener, C.T. Driscoll, D.A. Gay, N. Basu, B.A. Monson, K.F. Lambert, H.A. Morrison, J.T. Morgan, K.A. Williams, A.G. Soehl, 2011, *Great Lakes Mercury Connections: The Extent and Effects of Mercury Pollution in the Great Lakes Region* (Gorham, Maine: Biodiversity Research Institute, 2011).

Invasive Species

- One survey respondent reported progress made by a project to improve control of sea lampreys, an invasive species in the Great Lakes, for which his organization had received $8,203,561 in GLRI funding. According to the U.S. Geological Survey, sea lampreys have devastated fish communities in the Great Lakes by feeding on the bodily fluids of host fish, such as the native lake trout. According to the respondent, the project worked to develop a new sea lamprey control technique that relies on the sea lamprey's sense of smell. The respondent reported that field trials using this technique increased trapping efficiencies by up to 53 percent and that traps used with this technique can capture more than twice as many sea lampreys as traps that do not use the technique. According to the respondent, this new technique may make it possible not only to capture more sea lampreys but also to reduce the amount of pesticides introduced into the Great Lakes that are used to kill sea lampreys.

Sea lampreys

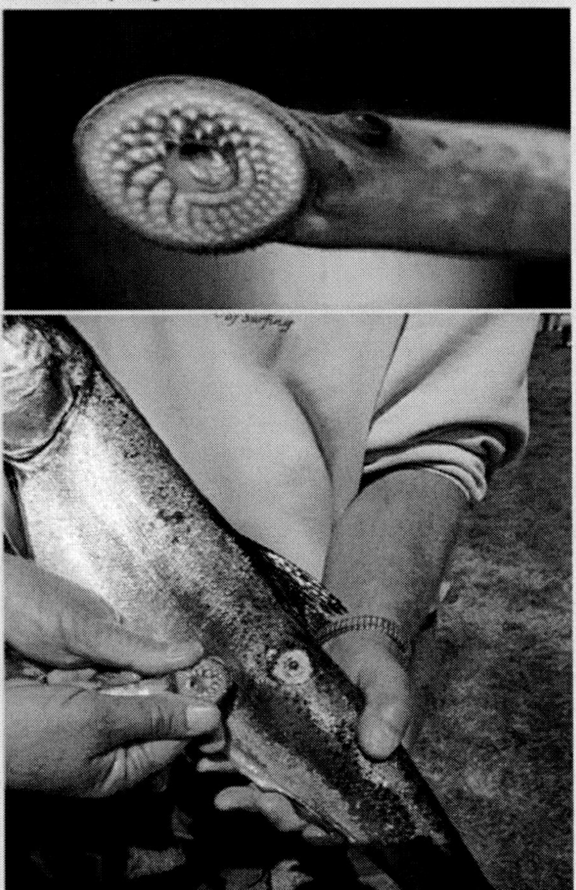

Source: Great Lakes Fishery Commission.

(Continued).

> Sea lampreys are parasites that attach to fish and feed on the fish's body fluids, often killing the fish. A single lamprey can kill up to about 40 pounds of fish in its lifetime. Sea lampreys prey on all species of Great Lakes fish, including economically important fish such as lake trout, salmon, whitefish, chubs, and walleye. Sea lampreys were a major cause of the collapse of lake trout, whitefish, and chub populations in the Great Lakes during the 1940s and 1950s. Sea lampreys are native to the Atlantic Ocean. They were observed in Lake Erie in 1921 following improvements to the canal that bypasses Niagara Falls and were observed in all five of the Great Lakes by 1939.

- Officials we interviewed from the U.S. Coast Guard told us that the agency's invasive species-related projects will increase its enforcement capability and address some technical problems relating to the installation and operation of freshwater ballast treatment systems for vessels operating on the Great Lakes. Ballast water is water that is taken into or discharged from a ship to accommodate changes in weight as a ship loads or unloads cargo. According to the U.S. Coast Guard, ballast treatment systems are intended to reduce the number of invasive species that can be transported into the Great Lakes in ballast water. The U.S. Coast Guard's Research and Development Center worked with the Naval Research Laboratory to design, build, test, and verify a system for testing treated ballast water under both shoreside and shipboard conditions. According to U.S. Coast Guard officials, this project will facilitate rigorous testing of the performance of ballast water treatment systems under shipboard conditions, ensuring greater protection against the introduction of invasive species into the Great Lakes.

Nearshore Health and Nonpoint Source Pollution

- One survey respondent reported that her organization received $85,000 in GLRI funding for a project to reduce the amount of bacteria in beach sand at a park on Lake Erie, in New York. According to the respondent, work completed in 2011 with GLRI funds determined that beach sand is acting as a reservoir for *Escherichia coli*, abbreviated as *E. coli*, which can negatively impact public health.[55] The organization then compared three sand grooming techniques to no beach grooming. According to the respondent, the results indicated that sand grooming using a specific tractor attachment can reduce the amount of bacteria in the sand. In 2012, the organization compared water quality when sand was groomed using two different grooming techniques. The respondent reported that preliminary results indicate that grooming the beach daily with the specific tractor attachment significantly reduced the number of beach closures. The respondent reported that this fiscal year 2011 to 2012 GLRI project resulted in an improvement in water quality.

- The Natural Resources Conservation Service received $75 million in GLRI funding from fiscal years 2010 to 2012 to target conservation efforts in selected priority areas. According to an agency official, this funding provides an opportunity to implement additional scientifically proven conservation practices in the priority watersheds, accelerating conservation practice implementation above what other programs would have provided. Landowners can take advantage of GLRI funds through a Natural Resources Conservation Service cost-share assistance program to install conservation practices. The agency helps landowners with conservation planning using various conservation practices, such as cover crops, conservation crop rotations, filter strips, prescribed grazing, and wetlands restoration. For example, the practice of cover crops establishes close-growing grasses, or other crops, to help improve soil and water quality by reducing soil erosion, among other things. A filter strip is a strip of herbaceous vegetation that is situated between crop or grazing land and a stream, river, or wetland, in order to reduce contaminated runoff. An agency official reported that GLRI funded efforts have led to the use of cover crops and filter strips on nearly 70,000 acres and 143 acres of priority watersheds respectively. For example, one landowner used the funds to develop a prescribed grazing system that encourages groundwater recharge, improves soil quality and prevents sediment and nutrient losses. The landowner is also installing a waste collection system to keep contaminated water out of nearby surface water and plans to plant 34 acres of cover crop.

Buffer zones and filter strips

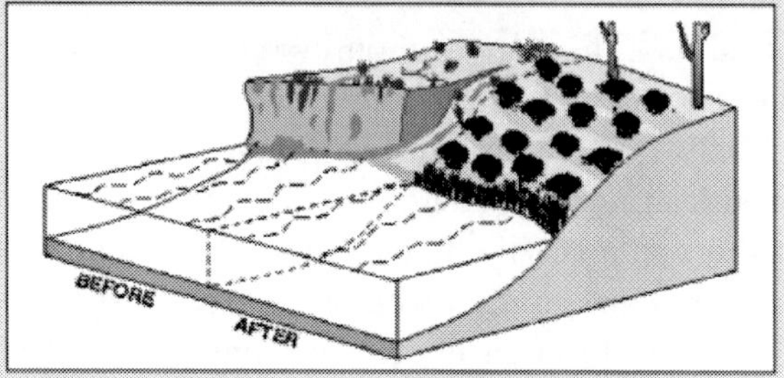

Source: Natural Resources Conservation Service.

Buffer zones and filter strips are conservation practices that farms can use to reduce erosion and nutrient risks. They are strips of natural vegetation along the banks of a stream, lake, or other water body that separate the water from developed areas. Buffers are important because they

- provide natural habitat for birds, mammals, and other wildlife;
- provide shade to keep stream water cooler, which can improve habitat for instream organisms such as fish and frogs;
- stabilize stream banks with their root systems and prevent erosion; and
- preserve water quality by capturing sediment and chemicals from runoff, and slow runoff to help prevent flooding.

Habitat and Wildlife Protection and Restoration

- A survey respondent reported that his organization had a project to help restore a marsh area that is among the roughly 15 costal marsh areas along Lake Michigan's Lower Peninsula shoreline. The organization received $783,823 in GLRI funds for the project. One component of the project was to replace seven misaligned culverts. According to the respondent, the new culverts will allow unimpeded fish passage to more than 12 miles of stream habitat, which improves habitat for trout and other aquatic organisms. In addition, the respondent reported that the project restored a mile-long section of trout stream back to its original watercourse. This was necessary because the hydrology of an adjacent 75-acre wetland had been significantly degraded when the trout stream had been diverted into a straightened channel decades ago. The project also resulted in the eradication of types of invasive plants. The survey respondent reported that this project will improve fisheries habitat, help restore hydrology to a large wetland for waterfowl and many other species of wildlife and fish, and improve public recreational opportunities.

Barriers to fish passage

Source: Ducks Unlimited.

(Continued).

> Barriers to fish passage, such as dams and misaligned culverts, result in habitat fragmentation and can impede the passage of fish species to and from their historic spawning grounds. According to a 2010 report by the Association of Fish and Wildlife Agencies, one-third of all Great Lakes fish are estimated to use tributaries as their principal spawning and nursery habitats and, in the Lake Michigan watershed alone, barriers to fish passage have reduced nearly 19,000 miles of available stream habitat to 3,300 miles. Reconnecting fragmented streams and rivers increases habitat connectivity and can create healthier aquatic ecosystems for fish.
>
> Note: National Fish Habitat Board, *Through a Fish's Eye: The Status of Fish Habitats in the United States 2010* (Washington, D.C.: Association of Fish and Wildlife Agencies, 2010), 20.

- Officials from the Fish and Wildlife Service told us that the agency has received $11,590,857 in GLRI funds in fiscal years 2010 to 2012 for a project that focuses on restoring habitats for native lake sturgeon, brook trout, migratory birds, and threatened and endangered species populations within the Great Lakes Basin by removing barriers to fish passage, stabilizing stream banks and riparian areas (narrow vegetated areas adjoining rivers, streams, and lakes), improving in-stream habitat and restoring costal, wetland and upland areas. According to agency officials, the project has protected, restored, or enhanced more than 2,000 acres of wetlands and uplands, removed or bypassed more than 30 fish passage barriers, and reopened more than 210 stream miles to fish movement. Agency officials said that improving aquatic connectivity in the Great Lakes Basin is one of the more prominent achievements of the project, and that removing barriers to fish passage creates a healthier aquatic habitat and improves water quality and sediment management. In addition, according to agency officials, free-flowing rivers provide new recreational opportunities.

Accountability, Education, Monitoring, Evaluation, Communication, and Partnerships

- One survey respondent reported that his organization received GLRI funding to map and describe all of the wastewater treatment plants in the Great Lakes Basin, both in the United States and Canada, and to develop a binational aquatic invasive species response plan. The respondent's organization received $300,000 in GLRI funding for these projects. According to the respondent, the project to map and describe wastewater treatment plants resulted in the first ever binational map of these facilities and will inform management decisions about the level and consistency of water treatment and help protect human and environmental health. In addition, the respondent reported that the aquatic invasive species response plan developed by his organization provides a foundation for work on a basinwide plan that will be an important backup plan in the event that aquatic invasive species prevention efforts fail. According to the respondent, the plan his organization developed will directly assist the governments of the United States and Canada in meeting their obligations under the 2012 amendment to the Great Lakes Water Quality Agreement, which requires the two governments to develop and implement an early detection and rapid response initiative for aquatic invasive species, among other things.
- Officials we interviewed from the National Oceanic and Atmospheric Administration told us that the agency's dissemination of training materials for educators and climate curricula for elementary through high school students, which is customized for the Great Lakes region, is on target to reach more than 3,000 students and 50 teachers by the end of fiscal year 2013. As a result, officials said, the number of institutions incorporating this information into existing curricula far exceeds the GLRI Action Plan targets. Specifically, the Action Plan's 2013 target for the measure that addresses the number of educational institutions incorporating Great Lakes protection and stewardship criteria into their environmental education curricula is 10 institutions. This measure corresponds to the Action Plan's goal to increase outreach and education for the Great Lakes so that students understand the benefits and ecosystem functions of the Great Lakes and are able to make decisions to ensure that restoration investments are enhanced over time.

In addition to these examples of progress, respondents to our survey and other stakeholders we interviewed reported that the GLRI had allowed them to do restoration work that they previously identified as important but could not undertake because of limited funds or staff, among other reasons. For example, one survey respondent noted that due to GLRI funding, progress has been made on problems that had been languishing for years due to inadequate funding such as high-priority restoration projects, aquatic invasive species control, and actions to reduce beneficial use impairments. In addition, a state agency official we interviewed told us that without GLRI funding his state would not have had the capacity to help coordinate restoration efforts for areas of concern or conduct necessary monitoring of fish toxicity. Furthermore, in commenting on a draft of this report, National Park Service officials noted that GLRI funds have proven to be invaluable for helping the Service accomplish significant restoration of wetlands and areas affected by invasive species, among other things.

Overall Great Lakes Restoration Progress Is Difficult to Quantify

We, and other organizations, have documented the difficulties associated with efforts to quantify overall progress toward Great Lakes restoration. For example, we reported in 2004 that it is difficult to describe restoration progress across the basin because of the unique environmental conditions of each of the Great Lakes.[56] Specifically, the Great Lakes are not one contiguous water body but rather distinct lakes with unique environmental conditions—such as lake depth—that present challenges to setting goals and developing a monitoring system that can be used to describe restoration progress across the basin and also capture the uniqueness of each lake. In addition, according to a 2006 United Nations report, it is usually difficult, and sometimes impossible, to attribute changes in the state of a large ecosystem solely to the efforts of a specific ecosystem management program. [57] Furthermore, several Task Force agency officials, nonfederal stakeholders, and subject matter experts told us that it may take time for some Great Lakes restoration efforts to show measurable results.

EPA officials told us that it may not be possible to summarize Great Lakes restoration progress with one simple statement due to the size and complexity of the Great Lakes ecosystem. They said that when the Action Plan was created, the Task Force agencies incorporated a measure the agency had created as part of its GPRA reporting requirements, the Great Lakes 40-point scale (also called the Great Lakes Index), as a measure of progress intended to represent progress across the focus areas by measuring improvement in the overall health of the aquatic ecosystem. The index rates eight indicators on a scale from 1 to 5, where 1 is poor, and 5 is good, to produce an assessment of overall Great Lakes health that is represented by one number.[58]

However, some stakeholders and subject matter experts we interviewed raised concerns about the index and some of its indicators. For example, they said that the index is missing key ecosystem indicators, such as those related to invasive species and the health of native species. In addition, they said that the index is better suited for assessments of longterm, as opposed to annual, changes in the health of the ecosystem. EPA officials also questioned how meaningful it was to use just one number to convey information about the state of the multiple ecosystems and indicators within the Great Lake basin.

EPA officials also explained that the GLRI is a relatively new undertaking and that getting visible, lake-wide and basin-wide ecosystem improvements from Great Lakes restoration efforts is a slow process. According to the officials, benefits, such as improvements in areas of concern, are being seen at the local level and by the end of the GLRI's lifespan in 2014, EPA will be able to show progress at a local level as a step toward improving the health of the lakes overall.

The Task Force agencies' draft framework includes direction to report on restoration progress and accomplishments, and transfer knowledge and lessons learned, among other things. The draft framework also notes that, while assembling project specific results can begin to offer a picture of cumulative progress and local success, reporting on GLRI at the program level can inform whether or not the GLRI is making restoration progress at other levels, such as the regional, lake, or ecosystem level.

Nonfederal Stakeholders Identified Key Factors That May Affect and May Limit GLRI Progress

In response to our survey, respondents reported on what they believed to be the key factors that may affect and may limit GLRI progress. They cited practical issues related to implementing GLRI projects, such as the time it takes to complete quality assurance requirements and obtain permits. They also cited broader issues outside the scope of the Action Plan, for which realistic solutions may be expensive, pointing in particular to inadequate infrastructure for wastewater and stormwater and the effects of climate change.

Factors Related to Implementing GLRI Projects Affected Progress

Survey respondents most often reported that their GLRI projects started later than planned because of quality assurance requirements and the length of time needed to obtain permits. It is important to emphasize that quality assurance requirements for environmental programs play a critical role in ensuring the success of those programs and that permits for environmental activities are required by a number of federal and state laws. Moreover, survey respondents did not suggest that either process should be eliminated when asked to comment on GLRI's quality assurance requirements. Nonetheless, later start times can slow GLRI implementation and, as a result, may affect the pace of progress.

- *Quality assurance requirements.* Organizations that receive GLRI funds through federal grants may be required to meet specific quality assurance requirements for projects and tasks involving environmental data in order to receive the funds.[59] Seventy-seven percent (135) of the 176 survey respondents reported that they needed to complete quality assurance requirements for at least one of their GLRI projects and, that on average, it took nearly 4 months to complete those requirements. Of those 135 respondents, 28 percent (38) reported that their projects started later than planned because of the time it took to complete quality assurance requirements, and 13 percent (17) reported that they could not start their project until the next spring because the ground or water was frozen. EPA officials told us that the GLRI quality assurance process was slow at the start of the program because the agency had not previously assessed quality assurance requirements for the kind of work being proposed in some of the projects. However, several survey respondents reported that the time it takes to complete GLRI quality assurance requirements has decreased since the program began in fiscal year 2010.
- *Time needed to obtain permits.* Organizations implementing GLRI projects might need to obtain federal or state permits for certain

> efforts, such as using herbicides to reduce invasive species, conducting controlled burning to enrich soil nutrients, or enhancing habitat for endangered species and communities. About half (87) of the 176 survey respondents reported that they needed a permit for at least one of their GLRI projects and that it took an average of 5 months to obtain the necessary permits. Of those 87 respondents, 31 reported that their projects started later than planned as a result of the time needed to obtain permits, and 13 reported that they could not start their project until the next spring because the ground or water was frozen. In addition, in the fiscal year 2011 GLRI progress report, the Task Force agencies identified permit processing as one of the factors for slowed project implementation for the habitat and wildlife protection and restoration measure of progress addressing the number of acres of coastal, upland, and island habitats protected, restored, and enhanced.

Survey respondents also reported that the timing of award notification resulted in later-than-planned start times for their GLRI projects (30 percent, or 52 out of 176, respondents),[60] and that weather events limited the implementation of their GLRI projects (26 percent, or 45 out of 176, respondents). Specifically, 16 percent (29) reported that weather events caused their GLRI projects to be completed later than planned, 15 percent (27) reported weather events caused a suspension of GLRI activities, and 6 percent (11) reported that weather events resulted in later-than-planned start times for their GLRI projects.[61] Weather events may cause GLRI projects to start later than planned because work cannot proceed on some projects—such as those that involve planting, doing construction, or monitoring stormwater—if there is a drought, too much rain, or the ground does not freeze in winter, among other things. For example, three survey respondents reported that work on their projects in wetlands was delayed because the wetlands involved with their organization's projects did not sufficiently freeze one winter.

Factors Outside the Scope of the Action Plan May Limit GLRI Progress

Once started, the success of GLRI projects and of the GLRI as a whole can depend upon factors outside the scope of the Action Plan—that is, factors that are not addressed by the Action Plan's long-term goals, objectives, or measures of progress—that affect the Great Lakes ecosystem. One such factor is atmospheric deposition, which is a process that transfers pollutants from the air to the earth's surface and can significantly impair water quality in the nation's rivers, lakes, bays, and estuaries, and harm human health and aquatic ecosystems.[62]

Airborne pollutants transferred by atmospheric deposition can fall to the ground in precipitation or as a gas or particle and be deposited either directly onto the surface of a water body or onto land and then transported into a water body through runoff. Atmospheric deposition contributed to the presence of toxaphene in the Great Lakes. Toxaphene is an insecticide that was once primarily used in southern states from 1947 to 1980, and it was banned in the United States for all uses in 1990. It is toxic for many aquatic organisms and accumulates in fish and shellfish causing health problems in humans who consume them.

We asked survey respondents to rank nine factors that are not addressed in the Action Plan on the basis of their potential to limit the restoration of the health of the Great Lakes ecosystem. As indicated in figure 5, survey respondents most frequently reported that inadequate infrastructure for wastewater or stormwater (72 percent, or 126 of 176, respondents) and the effects of climate change (65 percent, or 114 of 176, respondents) have the greatest potential to limit the restoration of the health of the Great Lakes' ecosystem in comparison with other factors outside the scope of the Action Plan.

Survey respondents also reported that factors such as the removal of water from the Great Lakes for use outside of the region and the effects of population growth in the Great Lakes region have the potential to limit the restoration of the health of the Great Lakes' ecosystem to a great or very great extent by 57 percent (101 out of 176 respondents) and 49 percent (87 out of 176 respondents) respectively.

In addition, we asked 21 Great Lakes subject matter experts we interviewed to rank these factors, and 13 of the 21 experts identified the same top two factors the survey respondents did, noting that both inadequate infrastructure for wastewater or stormwater and the effects of climate change have the potential to limit the restoration of the health of the Great Lakes ecosystem to a great or very great extent.

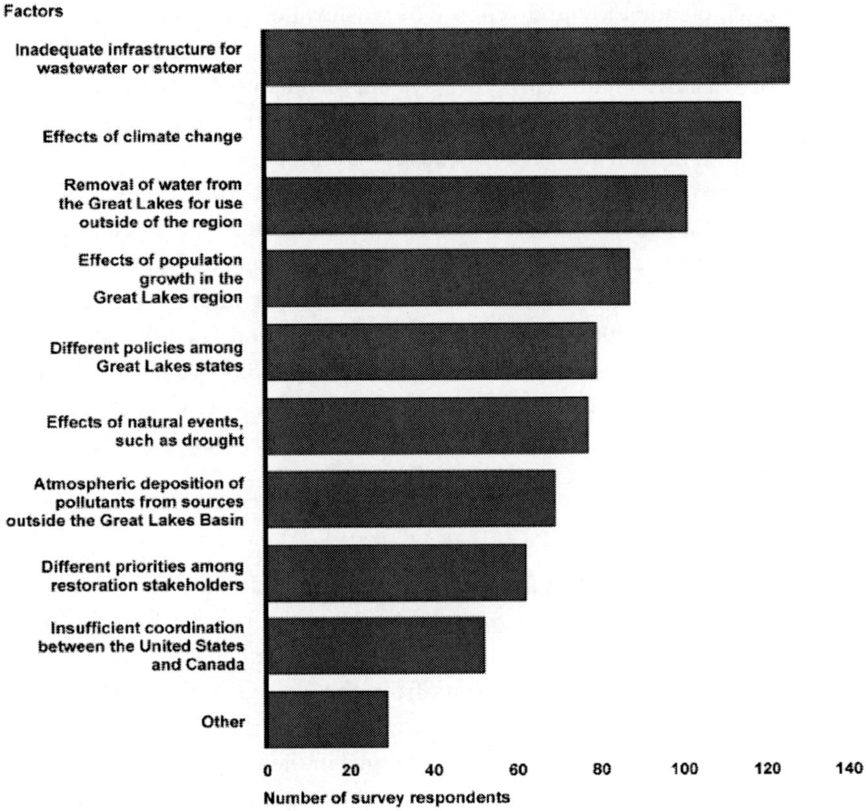

Source: GAO analysis of survey responses.

Figure 6. Survey Respondents Rated Factors Outside the Scope of the Great Lakes Restoration Initiative Action Plan That Have the Potential to Limit the Restoration of the Health of the Great Lakes Ecosystem to a Great or Very Great Extent.

Infrastructure

Inadequate infrastructure for wastewater or stormwater refers to the deteriorating condition of the nation's wastewater systems. These systems include sewer pipes that convey wastewater from homes and businesses to treatment facilities before discharging it into water bodies or land. In a 2004 report to Congress, EPA estimated that more than 1.2 million miles of pipes delivered wastewater into these systems. EPA has also reported that the vast majority of the pipes making up the nation's wastewater systems were installed more than 50 years ago and are reaching the end of their useful lives. In addition to the age of these systems, their conveyance capacity (i.e., the rate at which they can transport water) is often exceeded during rainfall or snowmelt, which adds to the discharge of wastewater into U.S. surface waters. According to the American Society of Civil Engineers, the age and inadequate capacity of wastewater infrastructure systems lead to the discharge billion of gallons of untreated wastewater into U.S. surface waters each year.[63]

Inadequate infrastructure can cause conditions that may slow, if not negate, restoration efforts, such as to improve water quality or reduce the number of days a beach is closed. This is because most wastewater systems are designed to discharge flows that exceed conveyance capacity directly to rivers, streams, and other surface waters. One Great Lakes interest group reported that five U.S. cities alone discharged 41 billion gallons of untreated sewage and stormwater into the Great Lakes from January 2009 to January 2010.[64] In its most recent clean water needs survey report, EPA stated that $19.7 billion was needed to address publicly owned wastewater and stormwater infrastructure needs in the Great Lakes Basin.[65]

Climate Change

Climate change is associated with increasing land and water temperatures and rising sea levels. There is uncertainty about the precise future effects of climate change on any particular region of the country, but a strong scientific consensus has emerged in recent years from the U.S. Global Change Research Program and the National Research Council that the future likely entails greater risks of flooding, drought, and changes in the frequency and severity of storms.[66] In addition, several organizations have predicted specific effects of climate change in the Great Lakes, which could negatively affect GLRI restoration efforts, such as to remove contaminated sediments from areas of concern or reduce the introduction of invasive species. For example, according to a 2000 report by the U.S. Global Change Research Program, drought could lead to certain situations in the Great Lakes that may require additional dredging of sediments at an annual cost of from $75 million to $125 million—about $101.6 million to $129.6 million in 2013 dollars—simply to maintain shipping lanes.[67] In addition, the U.S. Global Change Research Program reported in 2009 that warming water temperatures can lead to increased numbers of aquatic invasive species, which tend to thrive under a wide range of environmental conditions, and a decline in native species, which are adapted to a narrower range of conditions.[68] As we noted previously, invasive species such as the zebra mussel have caused extensive ecological and economic damage to the Great Lakes.

One way to reduce the potential effects of climate change is to invest in enhancing resilience. As defined by the National Academies, resilience is the ability to prepare and plan for, absorb, recover from, and more successfully adapt to adverse events. 69 Enhanced resilience results from better planning to reduce losses, rather than waiting for an event to occur and paying for recovery afterward. Many jurisdictions in the Great Lakes region have recognized the threat of climate change and committed resources to prepare for its effects, according to a 2012 report focused on the region.[70] Among these efforts, several Great Lakes states and the city of Chicago have each created climate change action plans that identify steps being taken to reduce the effects of climate change. For example, the city of Chicago identified the use of green roofs as one of many climate change mitigation and adaptation actions in its 2008 *Chicago Climate Action Plan*.[71] In its subsequent progress report, Chicago reported that more than 4 million square feet of green roofs had been planned or completed since 2008.[72] A green roof is a vegetative layer grown on a rooftop that provides shade and removes heat from the air, reducing temperatures of the roof surface and the surrounding air. As a result, green roofs act as insulators for buildings and reduce the energy needed to provide cooling and heating, which can decrease the production of associated air pollution and greenhouse gas emissions that contribute to climate change.

In addition, some Great Lakes jurisdictions have recognized that climate change can exacerbate other factors that may limit Great Lakes restoration. As we reported in April 2013, 73 the Milwaukee Metropolitan Sewerage District officials employed what they called "green infrastructure" programs to make the district's sewer system more resilient to climate change by capturing and holding or slowing the flow of stormwater. This was done as part of broader efforts to meet growing demand for sewer capacity, and officials plan to incorporate climate change adaptation into infrastructure planning and design where it makes sense as their facilities age and are replaced over time.

Action Plan Does Not Clearly Account for Key Factors That May Limit Its Success and Overall Great Lakes Restoration Progress

The Action Plan does not include goals, objectives, or measures of progress that show how the GLRI should address inadequate infrastructure, the effects of climate change, or other key factors that may limit the Action Plan's success and overall progress concerning the health of the Great Lakes ecosystem, with two exceptions.[74] In some cases, such as with inadequate infrastructure for wastewater or stormwater and atmospheric deposition of pollutants from sources outside the Great Lakes Basin, the Action Plan notes that some of these issues are addressed by other federal programs. For example, the Action Plan states that infrastructure needs will be addressed through increased funding for EPA's Clean Water State Revolving Funds, a program that provides states and local communities with independent and sustainable sources of financial assistance, such as low- or no-interest loans, to fund water quality projects identified by the states and localities. EPA reported in 2010 that the program has provided more than $74 billion for these projects since 1988.[75] As we testified in March 2013, however, wastewater infrastructure needs are made more daunting by the limited resources and budgets facing all levels of government.[76]

The Action Plan acknowledges that the effects of climate change may have implications across all five of the Action Plan's focus areas, and that the needs of communities to adapt to those effects will be assessed and addressed by GLRI projects and programs where appropriate. However, the Action Plan does not state how these effects will be addressed. To date, EPA has made some modest efforts to acknowledge the issue in the context of the Great Lakes. In its 2012 request for grant applications, for example, EPA included $200,000 for two to five grants for projects to increase climate change resiliency in Great Lakes communities. EPA also included an incentive in the request for applications for 2012 for the applicants to consider climate change impacts through vulnerability assessments or the integration of climate change adaptation measures into their project. Specifically, applicants could increase their eligibility for a grant by addressing climate change in their proposal.

The GLRI was created to accelerate the protection, maintenance, and restoration of the integrity of the Great Lakes. If the Action Plan does not state how the GLRI should address key factors—such as inadequate infrastructure and the effects of climate change—then the ability of restoration efforts guided by the Action Plan to achieve the GLRI's goal is uncertain because these factors may negatively affect restoration efforts. EPA has acknowledged that the effects of climate change can be detrimental to the Great Lakes and stated in its fiscal year 2013 budget justification that it is imperative that consideration of climate change and climate adaptation be integrated into GLRI grants and projects to ensure the overall success of the GLRI.[77] Nonetheless, to date, EPA has made few connections between this key issue and the potential effects of climate change. At a series of public meetings and webinars beginning in May 2013, EPA requested input regarding whether the connection between the Action Plan focus areas and the protection of the Great Lakes from the effects of climate change should be expressed more clearly in the next Action Plan. In commenting on a draft of this report, EPA officials told us that they have heard from many stakeholders about the need for the fiscal year 2015 to 2019 Action Plan to place additional emphasis on climate change, and that there is broad agreement on this point among the task force agencies.

CONCLUSION

As directed by Congress, EPA and the other Task Force agencies began implementing the GLRI in fiscal year 2010 to address the stresses that threaten the health of the Great Lakes ecosystem and help restore the health of the Great Lakes and issued the GLRI Action Plan to guide implementation of the approximately $1.3 billion appropriated to the program. A collaborative and comprehensive approach, it is clear that both federal and nonfederal GLRI stakeholders believe that the program is making strides toward its goals. For example, GLRI efforts enabled EPA to remove the Presque Isle Bay area of concern from the list of areas of concern identified by the United States and Canada. Nonetheless, the threats to the Great Lakes persist, and EPA and the other Task Force agencies face significant challenges in ensuring the future success of the GLRI due to issues involving their abilities to assess and achieve progress. EPA and the Task Force agencies are currently obtaining input for the next version of the GLRI Action Plan for fiscal years 2015 to 2019.

EPA has assessed GLRI progress primarily by using information from various sources to determine whether the GLRI is meeting the annual targets for the 28 measures in the Action Plan. Assessments of GLRI progress could help GLRI stakeholders, Congress, and the public discern the extent to which the health of the Great Lakes ecosystem has been restored, as well as what has been achieved by the approximately $1.3 billion appropriated to the program since fiscal year 2010. Moreover, such assessments can be used by EPA and the Task Force agencies to assess the effectiveness of GLRI actions and adjust their efforts if needed. However, we believe EPA is not well positioned to produce comprehensive and useful assessments for several reasons. Specifically:

- Some of the Action Plan's long-term goals and objectives do not have measures of progress that link to them. We believe that clear linkages between a plan's goals, objectives, and measures are critical to achieving and assessing progress over time. Without such linkages in the current Action Plan, it is unclear how EPA will be able to assess progress in meeting the Action Plan's long-term goals and objectives.

- Similarly, without identifying linkages between long-term goals, objectives, and measures in the Action Plan for 2015 to 2019 now under development, it is unclear how EPA will be able assess progress in meeting the long-term goals and objectives in that plan.
- The measures of progress do not capture the results of many of the GLRI projects because more than half of the projects are not associated with an Action Plan measure of progress in GLAS, which is a mechanism EPA created for collecting information to monitor GLRI projects and progress. We recognize that some projects do not directly provide data for any of the Action Plan's 28 measures, and that the cumulative results of some of the projects without assigned measures may ultimately be captured by some of the existing measures. With so many instances in which a project does not have an assigned measure, it is unclear if EPA will have the information it needs to capture the results of these projects and, therefore, if the agency can assess the full extent of progress being made.
- GLRI projects may directly address multiple measures of progress, but GLAS limits users to reporting the information for just one measure for each project even if more measures are being addressed. Without complete progress information on GLRI projects, the data EPA is using for certain measures of progress cannot be complete, and EPA may be underreporting GLRI progress. EPA officials told us that they will consider addressing this limitation in GLAS, but they did not specify a time frame for doing so.
- Some Action Plan measures or targets may not be useful because, among other things, they track activities that may not be sufficient to lead to the desired GLRI goals. Without useful measures, EPA may not be able to determine that GLRI efforts are producing the desired results.
- EPA and the Task Force agencies have not fully established an adaptive management plan. Although there is no requirement that the Task Force agencies establish an adaptive management plan for the GLRI, EPA and several of the Task Force agencies agreed in a 2000 to incorporate adaptive management principles into their programs. EPA and the Task Force agencies recently issued a plan that includes most but not all of the key elements of adaptive management, and it does not provide details on how the elements of adaptive management will be implemented. Without a fully established adaptive management plan, EPA and the other Task Force agencies may be

limited in their ability to assess the effectiveness of GLRI efforts and adjust future actions to achieve the goals and objectives of the GLRI as results and ecosystem processes become better understood.

We recognize the potentially significant contributions that individual GLRI projects can make to resolving specific environmental and public health stresses that threaten the Great Lakes. However, we believe that these contributions need to be viewed in the context of larger factors affecting the Great Lakes, in order to more fully appreciate the long-term future of the Great Lakes and to develop optimal strategies to help ensure the best possible outcome. Some of these factors can lead to problems that will be very expensive to address. Most notably, inadequate infrastructure for wastewater and stormwater and the effects of climate change may lead to conditions that can negatively affect GLRI restoration efforts. One way to reduce the potential impacts of climate change is to invest in planning to reduce losses, rather than waiting for an event to occur and paying for recovery afterward. The Action Plan touches on these factors, but it does not directly address how these factors may affect GLRI efforts to restore the health of the Great Lakes ecosystem or provide strategies to address them. Without more clearly expressing that connection in the next Action Plan, EPA will not be able to help address the effects of these factors, including climate change, on GLRI restoration efforts.

RECOMMENDATIONS FOR EXECUTIVE ACTION

To address challenges the Task Force faces in producing comprehensive and useful assessments of progress and addressing factors that may limit GLRI progress, we recommend that the EPA Administrator, in coordination with the Task Force, as appropriate, take the following seven actions:

- ensure progress toward long-term goals or objectives that are identified in the Action Plan, but which do not have measures that link to them, is assessed;
- ensure that linkages between long-term goals, objectives, and measures are identified in the Action Plan for 2015 to 2019;
- ensure that the progress being made by projects that do not have an Action Plan measure assigned to them is captured in assessments of GLRI progress;

- capture complete information about progress for each of the measures that are addressed by a project;
- further evaluate the usefulness of the current measures and targets and the need, if any, for the creation of additional measures;
- establish an adaptive management plan that includes all of the key elements of adaptive management and provides details on how these elements will be implemented; and
- address how factors outside of the scope of the Action Plan that may limit progress, such as inadequate infrastructure for wastewater or stormwater and the effects of climate change, may affect GLRI efforts to restore the Great Lakes.

AGENCY COMMENTS AND OUR EVALUATION

We provided a draft of this report to EPA, the Departments of Agriculture, Commerce, Defense, Health and Human Services, Homeland Security, the Interior, State, and Transportation, and the Council on Environmental Quality for comment. In written comments from the EPA Region 5 Administrator, EPA generally agreed with the conclusions and recommendations in our report. EPA noted in particular that our report is well-timed because it comes at the beginning of a new 5-year GLRI planning cycle. The agency also noted that we rightly acknowledged the complexity of assessing and quantifying overall Great Lakes health, the difficulty of attributing specific environmental changes to specific projects or programs, and the slow pace of ecosystem change.

In addition, EPA noted that the agency is already taking actions consistent with our recommendations, including working to develop more appropriate measures of progress in the Action Plan for fiscal years 2015 to 2019 and finalizing the draft framework. The agency also noted it is working with the task force agencies to link all activities described in interagency agreements to goals, objectives, and measures of progress in the Action Plan to help to track projects where there is not currently an associated measure of progress for which data can be reported in GLAS. EPA also noted that the agency has heard from many stakeholders about the need for the next Action Plan to place additional emphasis on climate change, and that there is broad agreement on this point from Task Force agencies. Taken together, we believe comprehensive actions along the lines stated in the EPA letter would constitute important steps forward, and we will look forward to seeing the progress the agency has made in taking these actions in the 2015 to 2019 Action Plan.

In written comments from the Department of the Interior's Assistant Secretary for Policy, Management, and Budget, Interior noted that information in our report is accurate and relevant, and the agency acknowledged the challenges in quantifying the progress toward restoration goals that we identified in our report. The Department of Agriculture provided technical comments but did not comment on our recommendations.

In addition to these written comments, the Council on Environmental Quality submitted an e-mail on September 13, 2013, that further emphasized our findings on the need for accountability. Specifically, this e-mail stated that GLAS limitations have led to an underreporting of progress for certain measures, so that they are not most accurately reflecting what benefits are accruing under the measures and projects implemented. EPA and the Departments of Agriculture, the Interior, and Transportation, provided technical comments that we incorporated as appropriate.

J. Alfredo Gómez
Director, Natural Resources and Environment

APPENDIX I: OBJECTIVES, SCOPE, AND METHODOLOGY

This appendix provides information on the scope of work and the methodology used to examine (1) how the Great Lakes Restoration Initiative (GLRI) is implemented by the Task Force agencies and other stakeholders; (2) the methods that the Environmental Protection Agency (EPA) has in place to assess GLRI progress; (3) the progress identified by the Task Force agencies and nonfederal stakeholders; and (4) the views of nonfederal stakeholders on factors, if any, that may affect or limit GLRI progress.

To examine how the GLRI is implemented by the Great Lakes Interagency Task Force (Task Force) agencies and other stakeholders we analyzed key documents and interviewed individuals involved with the GLRI.[1] Specifically, we analyzed the *Great Lakes Restoration Initiative Action Plan Fiscal Years 2010-2014* (Action Plan) to understand its structure and identify the long-term goals, objectives, measures of progress and related annual targets, and principal actions in each of the Action Plan's five focus areas; [2] EPA GLRI financial reports and the agency's fiscal year 2014 budget justification to determine the amount of GLRI funds that have been allocated for fiscal years

2010 to 2013 to each Task Force agency and each focus area; and, information from the Great Lakes Accountability System (GLAS) to identify the number of GLRI projects in each focus area and to learn more about GLRI projects, such as location, size, and funding.[3] We also reviewed interagency agreements between EPA and the Task Force agencies that identify the amount of GLRI funds EPA will allocate to each agency and the GLRI projects the agencies will implement. In addition, we conducted interviews with federal stakeholders, officials from EPA headquarters, EPA's office responsible for implementing the GLRI, the Great Lakes National Program Office, and the other Task Force agencies. We also interviewed representatives from nonfederal GLRI stakeholders—such as the Nature Conservancy and the Great Lakes Indian Fish and Wildlife Commission[4]—and a prominent Great Lakes interest group, Healing Our Waters – Great Lakes Coalition, to gain an understanding of the GLRI and how it is implemented.[5] Nonfederal stakeholders are state and local governments, tribes, nongovernmental organizations, and academic institutions that have received GLRI funds. We also attended the Eighth Annual Great Lakes Restoration Conference in Cleveland in September 2012 to learn more about the GLRI and Great Lakes restoration, and webinars to learn about the GLRI grant application process and how to use GLAS.

To examine the methods that EPA has in place to assess GLRI progress, we analyzed the Action Plan, documentation about GLAS, information from GLAS about GLRI project measures of progress, [6] information provided by EPA on the sources of data for each of the plan's measures of progress, the draft *GLRI Adaptive Science-Based Framework for Great Lakes Restoration* (draft Framework), and EPA's Science Advisory Board's 2012 review of the Action Plan.[7] We interviewed EPA officials to discuss their methods of assessing GLRI progress. We used the interviews with Task Force agency officials described above, and we conducted interviews with relevant officials from each of the Great Lakes states—Illinois, Indiana, Michigan, Minnesota, New York, Ohio, Pennsylvania, and Wisconsin—to discuss the Action Plan. We also conducted interviews with subject matter experts, to obtain their views about the achievability and usefulness of the Action Plan's measures of progress and related annual targets.

As part of our analysis of the Action Plan, we evaluated the extent to which the long-term goals and objectives in the plan have corresponding measures of progress. One analyst first compared the objectives for each focus area with the goals. The analyst considered whether it appeared that an objective would clearly contribute to attaining a goal. For example, in the invasive species focus area, the objective of developing and refining/piloting technologies that contain or control invasive species appears to clearly contribute to attaining the goal of preventing the spread of invasive species. The analyst also considered whether there was consistent or similar language in any of the goals and objectives that helped identify links between them. For example, the analysts examined if aspects of a goal in one focus area could be identified in an objective in the same focus area, such as in the toxic substances and areas of concern focus area, we identified similar language between the objective of reducing the concentrations of polychlorinated biphenyls (PCBs) in fish through 2014 and the goal of reducing environmental levels of toxic chemicals to the point that all restrictions on the consumption of Great Lakes fish can be lifted. Objectives could correspond to multiple goals. The analyst then compared the measures for each focus area to the objectives and looked for consistent or similar language in any of them that helped identify links between the objectives and measures. The analyst considered whether a measure was closely related to an objective and whether it appeared that progress toward at least part of the objective could clearly be assessed by a corresponding measure of progress. Measures could correspond to multiple objectives. We requested additional information from EPA officials on goals, objectives, and measures as needed to improve our understanding of them. A second analyst examined the conclusions of the first. In cases where there were initially disagreements between the two analysts regarding the identification of linkages between goals, objectives, and measures, all disagreements were resolved through analyst discussions. Ultimately, there was 100 percent agreement between the analysts. In addition, we evaluated the extent to which EPA and the Task Force agencies had established an adaptive management plan for the GLRI by first reviewing guidance from several Task Force agencies to identify key elements of adaptive management. We then examined the extent to which the GLRI Action Plan and draft Framework addressed those elements.

For our interviews with officials from each of the Great Lakes states, we developed a set list of open-ended questions for these interviews to obtain official's views about the Action Plan, among other things. We identified the state agencies that we would contact by starting with the list of state agencies identified at EPA's Great Lakes website as interested parties.[8] We then contacted officials from each interested party to determine if their agency was the lead GLRI agency for the state or to learn if we should contact a different agency. As a result of this process, we identified the following as lead GLRI state agencies: the Illinois Department of Natural Resources; the Indiana

Department of Environmental Management; the Michigan Department of Environmental Quality; the Minnesota Pollution Control Agency; the New York Department of Environmental Conservation; the Ohio Lake Erie Commission; the Pennsylvania Department of Environmental Protection; and, the Wisconsin Department of Natural Resources.

For the interviews with subject matter experts, we developed a set list of questions about the Action Plan's long-term goals, objectives, and measures of progress. The experts we interviewed included state agency officials, members of academia, and representatives of nongovernmental organizations who had expertise in one or more of the Action Plan's five focus areas. We identified these experts by asking for recommendations of at least three experts in each of the focus areas from 24 representatives of nonfederal Great Lakes stakeholders or Great Lakes interest groups organizations: 12 were nonfederal attendees of the Eighth Annual Great Lakes Restoration Conference that we selected randomly to conduct interviews about the Action Plan and the GLRI, and 12 were representatives of nonfederal organizations that we contacted as part of learning more about the GLRI and the Great Lakes.[9] We received responses from 17 of the 24 individuals who recommended a total of 187 subject matter experts.[10] We then created a list of the 33 experts who had been recommended two or more times and invited 31 of them to participate in interviews to obtain their views on the Action Plan goals, objectives, and measures of progress, among other things.[11] Of those 31 experts, 21 agreed to be interviewed: five experts in focus area one; six experts in focus area two; three experts in focus area three; four experts in focus area four; and three experts in focus area five. Because we used a nonprobability sample, the information obtained from these interviews is not generalizable to other individuals with expertise in the Action Plan focus areas but provides illustrative information.

To examine the progress identified by the Task Force agencies and nonfederal stakeholders, we took several steps: we reviewed the fiscal year 2010 and 2011 GLRI progress reports that EPA and the other Task Force agencies issued to Congress in April and March 2013, respectively; 12 we administered a web-based survey to each of the 205 nonfederal stakeholders that, as of October 2012, had received GLRI funds from a Task Force agency and had a project identified in GLAS; we used the interviews with officials from each Task Force agency identified above; and, we visited several GLRI projects in Illinois.

We conducted a web-based survey of all of the nonfederal GLRI stakeholders that have reported projects in GLAS, as of October 10, 2012, to identify examples of GLRI progress by obtaining their views on how their GLRI projects have benefited the ecosystem, among other things. The questionnaire used for this study is in appendix II. EPA provided us with contact information for each of those stakeholders, and we sent at least one e-mail to a point of contact from each stakeholder to identify the best point of contact for our survey. We surveyed the universe of the 205 nonfederal stakeholders that have projects in GLAS, as of October 10, 2012, and received responses from 176 nonfederal GLRI stakeholders for a response rate of 86 percent.

We designed draft questionnaires in close collaboration with a GAO social science survey specialist. We conducted pretests with three nonfederal GLRI stakeholders to help further refine our questions, develop new questions, and clarify any ambiguous portions of the survey. We conducted the pretests over the phone, and we selected one GLRI nonfederal stakeholder that had received GLRI funds for one grant and two GLRI nonfederal stakeholders that had each received GLRI funds for multiple grants for the pretest.

We developed and administered the web-based questionnaire accessible through a secure server, and we e-mailed unique identification numbers and passwords to points of contact at the 205 nonfederal GLRI stakeholders February 4, 2013. We sent follow-up e-mail messages beginning February 11, 2013. Then we contacted all remaining nonrespondents by telephone, starting February 26, 2013. The questionnaire was available online until March 28, 2013. For questions that should have been skipped but were not, we attempted to contact the respondents for clarification and edited their responses where warranted.

Because this was not a sample survey, it has no sampling errors. However, the practical difficulties of conducting any survey may introduce errors, known as nonsampling errors. For example, difficulties in interpreting a particular question, sources of information available to respondents, or when entering data into a questionnaire or analyzing the data can introduce unwanted variability into the survey results. We took steps in developing our questionnaire, collecting the data, and analyzing it to minimize these errors. In addition, as indicated above, social science survey specialists designed the questionnaire in collaboration with GAO staff that had subject matter expertise. We then conducted three pretests to check that (1) the questions were clear and unambiguous, (2) terminology was used correctly, (3) the questionnaire did not place an undue burden on respondents, (4) the information could feasibly be obtained, and (5) the survey was comprehensive and unbiased. We made multiple contact attempts with nonrespondents during the survey by email, and some nonrespondents were also contacted by telephone. When we analyzed the data, an independent analyst checked all computer programs. Since this was a web-based survey, respondents entered their answers directly into the electronic questionnaire, eliminating the need to key data into a database, minimizing error.

We conducted a computer-enabled content analysis to analyze responses to the question "Please provide examples of how one or more of your organization's GLRI projects have benefitted the Great Lakes ecosystem." Of the 176 respondents, 163 provided information in response to this question. We categorized the responses along two dimensions: direct and indirect benefits, and observed or expected benefits. We defined these categories based on whether respondents reported that their organization's GLRI projects resulted in changes in the ecosystem, such as reduced beach closure days, or other types of changes, such as improved understanding of the ecosystem, and if respondents reported that they had observed the result or expected to observe them a later time. One reviewer developed content categories based on survey responses and, after obtaining agreement on the categories from the second reviewer, assessed and coded each survey response into those categories. The second reviewer examined the coding. In cases where disagreements among the two reviewers regarding the coding of responses into content categories were found, all disagreements were resolved through reviewer discussion. Ultimately, there was 100 percent agreement between the reviewers. In addition, we conducted follow-up interviews by phone and e-mail with those survey respondents whose examples we wanted to include in the report to collect clarifying information, if necessary.

In addition to the survey, we used the interviews with Task Force agency officials identified above to obtain information about GLRI progress. We also visited several GLRI projects in Illinois. We traveled to Illinois because EPA's Great Lakes National Program Office is located in Chicago. We met with representatives from two nonfederal GLRI stakeholders who took us to the sites of their GLRI projects; the two nonfederal GLRI stakeholders—the Chicago Park District and Waukegan Harbor Citizens Advisory Group—were recommended to us by an official from the Illinois Department of Natural Resources. To examine the views of nonfederal stakeholders on factors, if any, that may affect or limit GLRI progress, we used the web-based survey described above. Specifically, we used the survey to obtain nonfederal GLRI stakeholder views about factors that may have limited their ability to carry out GLRI projects and factors that have the potential to affect the restoration of Great Lakes health. As part of creating the survey, we developed lists of factors that might limit GLRI project implementation progress and factors that might limit the restoration of Great Lakes health in general through interviews we conducted with 19 randomly selected individuals at the Eight Annual Great Lakes Restoration Conference and interviews we conducted with representatives from five nonfederal GLRI stakeholders and one Great Lakes interest group. We narrowed the list of factors that might limit GLRI project implementation progress to four factors most often cited by interviewees as hindering their abilities to start a project or to successfully carry out a project, and asked survey respondents if they had experienced these factors and about the effects of each of these factors on their GLRI projects. We narrowed the list of factors that might limit the restoration of Great Lakes health in general to the nine factors most often cited by grouping similar factors together and eliminating those factors that are addressed by the Action Plan either through Action Plan long-term goals, objectives, or measures of progress. We eliminated those factors addressed by the Action Plan because this meant they had already been identified by the GLRI as factors that had the potential to limit restoration of Great Lakes health. We asked survey respondents to identify the extent to which each of these factors had the potential to limit the restoration of the health of the Great Lakes ecosystem.

We also used the interviews with subject matter experts mentioned above to obtain information about the factors that may limit progress by asking them to identify the extent to which each of the same nine factors we included in the survey had the potential to limit the restoration of the health of the Great Lakes ecosystem. We conducted this performance audit from June 2012 to September 2013 in accordance with generally accepted government auditing standards. Those standards require that we plan and perform the audit to obtain sufficient, appropriate evidence to provide a reasonable basis for our findings and conclusions based on our audit objectives. We believe that the evidence obtained provides a reasonable basis for our findings and conclusions based on our audit objectives.

APPENDIX II: SURVEY QUESTIONS

We conducted a web-based survey of nonfederal GLRI stakeholders that have reported projects in the Great Lakes Accountability System, as of October 10, 2012, using all of the questions below as stated here.[1] Nonfederal stakeholders are state and local governments, tribes, nongovernmental organizations, and academic institutions that that have received GLRI funds.

Survey of Great Lakes Restoration Initiative (GLRI) Grantees

United States Government Accountability Office

Introduction

The U.S. Government Accountability Office (GAO), an agency of the U.S. Congress, is conducting a review of the Great Lakes Restoration Initiative (GLRI). This review has been requested by the Chairman of the House Subcommittee on Water Resources and Environment, Committee on Transportation and Infrastructure. One of the topics we plan to address is how GLRI stakeholders characterize the extent of GLRI progress made and the factors that may limit progress. As part of addressing this topic, we are gathering information from GLRI grantees about their experiences implementing GLRI projects, strengths of the GLRI program, suggestions for improvements to the GLRI program, and Great Lakes restoration in general.

Your responses to the following questions will help us inform Congress about what is happening on the ground. We will aggregate the information that we gather through this survey in our report to Congress, and we do not plan to identify respondents beyond the type of the organizations; that is, we plan to discuss information collected from nongovernmental organizations, state agencies, tribes, academic institutions, and others. This report will be made available to the public.

To learn more about completing the survey, printing your responses, and whom to contact if you have questions, click here for help.

We appreciate you taking the time to provide your input on this.

Section 1 - Background

1. Please provide the following information for the primary person completing this questionnaire in the event we need to contact you to clarify a response.

Name:	
Title:	
Agency/Organization:	
E-mail:	
Telephone *(include area code)*:	

GLRI Projects Identified in the Great Lakes Accountability System for your Organization Funded in Federal Fiscal Years 2010, 2011, or 2012

2. In the Great Lakes Accountability System, your organization is identified as having ______ GLRI projects funded in federal fiscal years 2010, 2011, or 2012. (See list of projects above.)

Is this number correct?

○ Yes
○ No

(If no) For how many GLRI projects has your organization received funding directly from federal agencies?	[ ] GLRI projects
Please identify any additions or deletions of projects from the list that we included above.	

The next set of questions in sections 2 through 5 pertain to your organization's ability to implement its GLRI projects. The focus will be on various factors that may have limited your organization's ability to carry out these projects. By limited, we mean that you were unable to implement your organization's GLRI projects as intended, such as a delayed start time or reduced scope. When answering these questions, please consider **all of the GLRI projects for which you received GLRI funds directly from a federal agency in federal fiscal years 2010, 2011, and 2012.**

Section 2 - Implementation of GLRI Projects - Quality Assurance Requirements

This section focuses on quality assurance requirements for your organization's GLRI projects.

3. Did your organization need to complete quality assurance requirements **for any** of its GLRI projects?

○ Yes
○ No

3a. (If yes) On average, approximately how long (in months) did it take for your organization to complete and submit the quality assurance requirements for its GLRI projects? ***(Enter number.)***

[ ] Months

3b. Did the time it took your organization to complete and submit the quality assurance requirements result in starting **any** of your projects later than originally planned?

○ Yes - *Continue with question 3c.*
○ No *(Click here to skip to question 3h)*

3c. (If yes) How many of your organization's projects were started later than originally planned due to the amount of time it took to complete and submit the quality assurance requirements? *(Enter number.)*

[ ] Projects

3d. Of these projects that started later than originally planned, what was the **longest amount of time (in months) that a project start was delayed**, and in what federal fiscal year was the grant for that project awarded? *(Enter number and year.)*

Longest time a project start was delayed	[ ] Months
Federal fiscal year the grant for that project was awarded	[none]

3e. Please identify which federal agency's or agencies' quality assurance requirements process resulted in starting **any** of your organization's projects later than originally planned.

[ ]

3f. As a result of any of the time it took your organization to complete and submit the quality assurance requirements, did your organization have to delay the start time of **any** of its GLRI projects until the following spring to wait for the ground or water to thaw?

- ○ Yes
- ○ No

3g. Has the time it has taken your organization to complete and submit quality assurance requirements decreased over time? To answer this question, please consider the time it has taken to complete and submit quality assurance requirements **for similar types of projects.**

- ○ Yes
- ○ No
- ○ Not applicable, our organization completed quality assurance requirements only once or for different types of projects.

3h. Please provide any comments you would like to make about your organization's experience with GLRI quality assurance requirements.

[ ]

Section 3 - Implementation of GLRI Projects - Obtaining Permits

This section focuses on obtaining permits for your organization's GLRI projects.

4. Did **any** of your organization's projects need a federal, state, or local permit?

- ○ Yes
- ○ No

4a. (If yes) On average, approximately how long (in months) did it take to obtain the necessary federal, state, or local permits for your organization's GLRI projects? *(Enter number.)*

[ ] Months

4b. Did the time it took to obtain the permit(s) necessary to implement your organization's GLRI projects result in starting any of your organization's projects later than originally planned?

- ○ Yes - *Continue with question 4c.*
- ○ No *(Click here to skip to question 4g)*

4c. (If yes) How many of your organization's projects were started later than originally planned due to the amount of time it took to obtain the necessary permits? *(Enter number.)*

[ ] Projects

4d. Of these projects that started later than originally planned, what was the **longest amount of time (in months) that a project start was delayed**, and in what federal fiscal year was the grant for that project awarded? *(Enter number and year.)*

Longest time a project start was delayed	[ ] Months
Federal fiscal year the grant for that project was awarded	[none]

4e. What type of permit caused the delay in implementing **any** of your organization's GLRI projects? ***(Select all that apply.)***

- ☐ Local permit
- ☐ State permit
- ☐ Federal permit

4f. As a result of any of the time it took to obtain necessary permits, did your organization have to delay the start time of **any** of its GLRI projects until the following spring to wait for the ground or water to thaw?

- ○ Yes
- ○ No

4g. Please provide any comments you would like to make about your experience obtaining permits for any of your organization's GLRI projects.

[ ]

Section 4 - Implementation of GLRI Projects - Timing of Award Notification

This section focuses on the timing of the award notification for your organization's GLRI projects.

5. Was the start date of any of your organization's GLRI projects delayed because of the timing of when it received notice that its GLRI projects would be funded (award notification)?

- Yes
- No

5a. (If yes) For the projects that were delayed due to the timing of the award notification, on average, approximately how long (in months) was the interval between the time that your organization had planned to start these GLRI projects and when you actually started these GLRI projects? *(Enter number.)*

[ ] Months

5b. How many of your organization's projects were started later than originally planned due to the timing of the award notification? *(Enter number.)*

[ ] Projects

5c. Of these projects that started later than originally planned, what was the **longest amount of time (in months) that a project start was delayed,** and in what federal fiscal year was the grant for that project awarded? *(Enter number and year.)*

Longest time a project start was delayed	[ ] Months
Federal fiscal year the grant for that project was awarded	[none]

5d. As a result of the timing of the award notification, did your organization have to delay the start time of **any** of its GLRI projects until the following spring to wait for the ground or water to thaw?

- Yes
- No

5e. If any of your organization's projects were delayed because of the timing of award notification for reasons unrelated to seasons (as described in question 5d), please describe how the timing of award notification caused a delayed start time for one of those projects.

6. Please provide any comments you would like to make about any aspect of the GLRI award notification process.

Section 5 - Implementation of GLRI Projects - Weather and Other Factors

This section focuses on the effect of weather and other factors on your organization's GLRI projects.

7. Did **weather events** limit your organization's ability to carry out **any** of its GLRI projects **in the following ways at any point during the project? Weather events** include, but are not limited to, storms or limited rainfall, **but do not include seasons**. *(Select one answer in each row.)*

	Yes	No	Not sure
a. A weather event caused a delayed start to a GLRI project	○	○	○
b. A weather event caused a suspension of GLRI project activities for a period of time	○	○	○
c. A weather event resulted in a delayed completion of a GLRI project	○	○	○

For item a - Please describe the weather event and how it caused a delayed start to a GLRI project:

For item b - Please describe the weather event and how it caused a suspension of GLRI project activities for a period of time:

For item c - Please describe the weather event and how it resulted in a delayed completion of a GLRI project:

8. Other than the factors covered in any of the previous questions on this questionnaire, were there any other factors that limited your organization's ability to implement **any** of its GLRI projects as intended?

○ Yes
○ No

8a. (If yes) Please describe these other factors and their effect on your organization's GLRI projects.

Section 6 - GLRI Program Benefits

This section focuses on the benefits obtained from the GLRI Program.

9. Please provide examples of things that have gone well in your organization's experience with the GLRI program. This can include, but should not be limited to, your experiences with or observations about the federal agencies administering the program as well as your own experience implementing the program.

10. Please provide examples of how one or more of your organization's GLRI projects have benefitted the community in or near the project area.

11. Please provide examples of how one or more of your organization's GLRI projects have benefitted the Great Lakes ecosystem. (For this question we are looking for a high-level description of project outcomes that we can use to illustrate how GLRI projects are helping to restore the health of the Great Lakes ecosystem. It is not necessary to provide the evidence linking your organization's GLRI project to the ecosystem benefit you have identified. We realize that benefits for many GLRI projects may not be observable or quantifiable at this time.)

Section 7 - Great Lakes Restoration

The two questions in this section pertain to factors outside of the scope of the GLRI Action Plan that have the potential to affect the broader goal of restoring the health of the Great Lakes ecosystem. In responding to these questions, please consider whether these factors may affect Great Lakes restoration overall and not just your organization's GLRI projects.

12. Based on your experience and given current policies affecting the Great Lakes region, to what extent do the following factors have the potential to limit the restoration of the health of the Great Lakes ecosystem? *(Select one answer in each row.)*

	To a very great extent	To a great extent	To a moderate extent	To some or little extent	To no extent	No basis to judge	Not applicable
a. Effects of climate change	○	○	○	○	○	○	○
b. Effects of population growth in the Great Lakes region	○	○	○	○	○	○	○
c. Atmospheric deposition of pollutants from sources outside the Great Lakes basin	○	○	○	○	○	○	○
d. Inadequate infrastructure for wastewater or stormwater	○	○	○	○	○	○	○
e. Effects of natural events, such as drought	○	○	○	○	○	○	○
f. Removal of water from the Great Lakes for use outside of the region	○	○	○	○	○	○	○
g. Different policies among Great Lakes states	○	○	○	○	○	○	○
h. Insufficient coordination between the United States and Canada	○	○	○	○	○	○	○
i. Different priorities among restoration stakeholders	○	○	○	○	○	○	○
j. Other	○	○	○	○	○	○	○

For item i - Please specify which stakeholders you have found to have different priorities:

For item j - Please specify the other factor(s) that may limit the restoration of the Great Lakes ecosystem:

13. Looking at the above list of factors outside of the scope of the GLRI Action Plan that have the potential to affect the broader goal of restoring the health of the Great Lakes ecosystem, which would you say are the top three in terms of having the greatest potential to limit the restoration of the health of the Great Lakes ecosystem?

*(Enter the **letter designations** (a-j) from question 12 above.)*

The first most limiting factor is: ☐

The second most limiting factor is: ☐

The third most limiting factor is: ☐

Section 8 - Improvements to the GLRI Program and Survey Completion Question

14. What suggestion or suggestions would you make to improve the GLRI?

15. Please provide any additional comments you may have about the GLRI or Great Lakes restoration in general.

16. **Are you ready to submit your final completed survey to GAO?**
(This is equivalent to mailing a completed paper survey to us. It tells us that your answers are official and final.)

- ○ Yes, my survey is complete - *To submit your final responses, please click on "Exit" below.*
- ◉ No, my survey is not yet complete - *To save your responses for later, please click on "Exit" below.*

You may view and print your completed survey by clicking on the Summary link in the menu to the left.

Thank you very much for your assistance.

Print

Exit

APPENDIX III: GREAT LAKES RESTORATION INITIATIVE ACTION PLAN LONG-TERM GOALS, OBJECTIVES, AND MEASURES OF PROGRESS FOR EACH OF THE FIVE FOCUS AREAS

The GLRI Action Plan for fiscal years 2010 to 2014 is organized by five focus areas that, according to the federal agencies responsible for implementing the GLRI, encompass the most significant environmental problems in the Great Lakes: (1) toxic substances and areas of concern; (2) invasive species; (3) nearshore health and nonpoint source pollution; (4)

habitat and wildlife protection and restoration; and (5) accountability, education, monitoring, evaluation, communication and partnerships. Each focus area includes several long-term goals to address these problems, as well as objectives to be completed within the 5 years of the plan and measures of progress to ensure that efforts are on track to meet the longterm goals.

Focus Area One, Toxic Substances and Areas of Concern

Long-Term Goals

- Goal 1: Areas of concern are cleaned up, restoring the areas and removing the beneficial use impairments.
- Goal 2: The release of toxic substances in toxic amounts is prevented and the release of any or all persistent toxic substances to the Great Lakes Basin ecosystem is virtually eliminated.
- Goal 3: Exposure to toxic substances from historically contaminated sources is significantly reduced through source reduction and other exposure reduction methods.
- Goal 4: Environmental levels of toxic chemicals are reduced to the point that all restrictions on the consumption of Great Lakes fish can be lifted.
- Goal 5: The health and integrity of wildlife populations and habitat are protected from adverse chemical and biological effects associated with the presence of toxic substances in the Great Lake Basin.

Objectives

- By 2014, delist five areas of concern.
- By 2014, 46 beneficial use impairments will be removed in areas of concern.
- By 2011, 15 million pounds of electronic waste and 15 million pills of unwanted medicines will be collected or their release will have been prevented.
- By 2014, 45 million pounds of e-waste, 45 million pills of unwanted medicines, and 4.5 million pounds of household hazardous waste in the Great Lakes Basin will have been collected or their release will have been prevented.
- By 2014, 9.4 million cubic yards of contaminated sediments will be remediated.

- Through 2014, an annual average of up to 5 percent annual decline will be maintained or improved for the trend (year 2000 and on) in average concentrations of PCBs in whole lake trout and walleye samples.

Measures of Progress

- Number of areas of concern in the Great Lakes where all management actions necessary for delisting have been implemented (cumulative).
- Area of concern beneficial use impairments removed (cumulative).[1]
- Beneficial use impairment delisting project starts at areas of concern (cumulative).[2]
- Cubic yards (in millions) of contaminated sediment remediated in the Great Lakes (cumulative).[3]
- Pollution (in pounds) collected through prevention and waste minimization projects in the Great Lakes Basin (cumulative).
- Cumulative percentage decline for the long term trend in average concentrations of PCBs in Great Lakes fish.[4]

Focus Area Two, Invasive Species

Long-Term Goals

- Goal 1: The introduction of new invasive species to the Great Lakes Basin ecosystem is eliminated, reflecting a “zero tolerance policy” toward invasives.
- Goal 2: The risk of introduction of species, which are imported for various uses, in the Great Lakes is minimized.
- Goal 3: The spread of invasive species, by means of recreational activities, connecting waterways, and other vectors, beyond their current range is prevented.
- Goal 4: A comprehensive program for detection and tracking newly identified invasive species in the Great Lakes is developed and provides up-to-date critical information needed by decision makers for evaluating potential rapid response actions.
- Goal 5: An effective, efficient and environmentally sound program of integrated pest management for invasive species is developed and implemented, including program functions of containment, eradication, control, and mitigation.

Objectives

- By 2011, eight state aquatic nuisance species management plans will be established or revised to include rapid response capabilities. By 2014, eight state-based, multiagency rapid response plans will be implemented and 22 mock exercises to practice responses carried out under those plans and/or actual response actions will be completed.
- Six technologies that prevent the introduction of invasive species and four technologies that either contain or control invasive species will be developed or refined and piloted by 2011. Ten technologies that prevent the introduction of invasive species and five technologies that either contain or control invasive species will be developed or refined and piloted by 2014.
- By 2011, methodology and protocols will be piloted for the coordinated monitoring methodology and shared protocols for basinwide invasive species surveillance. By 2014, a basinwide surveillance program with shared sampling protocols and methodologies to provide early detection of nonnative species will be operational.
- By 2014, a 40 percent reduction in the yearly average rate of invasive species newly detected in the Great Lakes ecosystem will be achieved, compared with the period 2000-2009.
- By 2014, invasive species populations within the Great Lakes ecosystem will have been controlled and reduced, as measured in populations controlled to a target level in 6,500 acres of managed area and by removing 5,000 pounds of invasive species from the Great Lakes ecosystem.
- By 2014, approximately 10 million recreation and resource users will be educated on best practices that prevent the introduction and spread of invasive species.

Measures of Progress

- Rate of nonnative species newly detected in the Great Lakes ecosystem.
- Acres managed for populations of invasive species controlled to a target level (cumulative).
- Number multiagency plans established, mock exercises to practice rapid responses carried out under those plans, and/or actual rapid response actions (cumulative).

- Number of recreation and resource users contacted on best practices that prevent the introduction and spread of invasive species (cumulative).

Focus Area Three, Nearshore Health and Nonpoint Source Pollution

Long-Term Goals

- Goal 1: Nearshore aquatic communities consist of healthy, self-sustaining plant and animal populations dominated by native and naturalized species.
- Goal 2: Land use, recreation, and economic activities are managed to ensure that nearshore aquatic, wetland, and upland habitats will sustain the health and function of natural communities.
- Goal 3: The presence of bacteria, viruses, pathogens, nuisance growths of plants or animals, objectionable taste or odors, or other risks to human health are reduced to levels in which water quality standards are met and beneficial uses attained to protect human use and enjoyment of the nearshore areas.
- Goal 4: High-quality bathing beach opportunities are maintained by eliminating impairments from bacterial, algal, and chemical contamination; effective monitoring for pathogens; effective modeling of environmental conditions, where appropriate; and timely communications to the public about beach health and daily swimming conditions.
- Goal 5: A significant reduction in soil erosion and the loading of sediments, nutrients, and pollutants into tributaries is achieved through greater implementation of practices that conserve soil and slow overland flow in agriculture, forestry and urban areas.
- Goal 6: High quality, timely and relevant information about the nearshore areas is readily available to assess progress and to inform enlightened decision making.

Objectives

- By 2010, EPA will compile and map the highest priority watersheds for implementation of targeted nonpoint source pollution control measures.

- By 2014, remediation, restoration and conservation actions in at least one targeted watershed in each Great Lake Basin will control erosion, reduce nutrient runoff from urban and agricultural sources, and improve habitat to protect nearshore aquatic resources.
- By 2014, a baseline will be established for total suspended solids loadings from targeted tributaries.
- By 2014, a measurable decrease will be achieved in soluble phosphorous loading from 2008 levels in targeted tributaries.
- By 2014, the causes of nutrient-related nearshore biological impairments will be better understood and, following local or watershed remedial actions, the number and severity of incidences of harmful algal blooms, avian botulism, and/or excessive *Cladophora* growth will be significantly reduced from 2008 levels.
- By 2014, a comprehensive nearshore monitoring program will have been established and implemented, including a publicly accessible reporting system, based on a suite of environmental indicators.
- By 2014, 50 percent of high-priority Great Lakes beaches will have been assessed using a standardized sanitary survey tool to identify sources of contamination.[5]
- By 2014, 20 percent of high-priority Great Lakes beaches will have begun to implement measures to control, manage or remediate pollution sources identified through the use of sanitary surveys.
- By 2014, rapid testing or predictive modeling methods (to improve the accuracy of decisions on beach postings to better protect public health) will be employed at 33 percent of high-priority beaches.
- By 2014, the area of agricultural lands in conservation and/or utilizing conservation tillage practices will increase by 50 percent over 2008 levels.

Measures of Progress

- Five-year average annual loadings of soluble reactive phosphorous from tributaries draining targeted watersheds.[6]
- Percentage of beaches meeting bacteria standards 95 percent or more of beach days.
- Extent (sq. miles) of Great Lakes harmful algal blooms.[7]
- Annual number of days U.S. Great Lakes beaches are closed or posted due to nuisance algae.[8]
- Annual volume of sediment deposition in defined harbor areas in targeted watersheds (cubic yards).[9]

- Acres in Great Lakes watershed with U.S. Department of Agriculture conservation practices implemented to reduce erosion, nutrients and/or pesticide loading under Farm Bill programs.[10]

Focus Area Four, Habitat and Wildlife Protection and Restoration

Long-Term Goals

- Goal 1: Protection and restoration of Great Lakes aquatic and terrestrial habitats, including physical, chemical, and biological processes and ecosystem functions, maintain or improve the conditions of native fish and wildlife.
- Goal 2: Critical management activities (such as stocking native fish and other aquatic species, restoring access of migratory fish species at fish passage barriers, and identifying and addressing diseases) protect and conserve important fish and wildlife populations.
- Goal 3: Sound decision making is facilitated by accessible, site specific and landscape-scale baseline status and trend information about fish and wildlife resources and their habitats.
- Goal 4: High-priority actions identified in strategic plans (such as state and federal species management, restoration and recovery plans, Lakewide Management Plans, Remedial Action Plans, and others) are implemented, lead to the achievement of plan goals, and reduce the loss of fish and wildlife and their habitats.
- Goal 5: Development activities are planned and implemented in ways that are sensitive to environmental considerations and compatible with fish and wildlife and their habitats.

Objectives

- By 2014, 4,500 miles of Great Lakes rivers and tributaries will be reopened and 450 barriers to fish passage will be removed or bypassed.
- By 2014, 82 percent of recovery actions for federally listed priority species will be implemented.
- By 2014, 53 percent of populations of native aquatic non-threatened and endangered species are self-sustaining.

- By 2014, 97,500 acres of wetlands, wetland-associated uplands, and high-priority coastal, upland, urban, and island habitats will be protected, restored or enhanced.
- By 2014, 100 percent of U.S. coastal wetlands in the Great Lakes Basin will be assessed.
- By 2014, 30 habitat-related beneficial use impairments will be delisted across the areas of concern.

Measures of Progress

- Miles of rivers reopened for fish passage.
- Number of fish passage barriers removed or bypassed.
- Number of species delisted due to recovery.
- Percentage of recovery actions implemented for priority listed species.
- Percentage of populations of native aquatic non-threatened and endangered species self-sustaining in the wild.
- Number of acres of wetlands and wetland-associated uplands protected, restored and enhanced.
- Number of acres of coastal, upland, and island habitats protected, restored and enhanced.
- Percent of U.S. coastal Great Lakes wetlands assessed.
- Number of habitat-related beneficial use impairments removed from the 27 U.S. areas of concern so impaired.[11]

Focus Area Five, Accountability, Education, Monitoring, Evaluation, Communication and Partnerships

Long-Term Goals

- Goal 1: A cooperative monitoring and observing system provides a comprehensive assessment of the Great Lakes ecosystem.
- Goal 2: The necessary technology and programmatic infrastructure supports monitoring and reporting, including Great Lakes Restoration Initiative project deliverables by all agencies and participating stakeholders. Data and information are provided in reports that are public friendly, timely and available on the Internet. Reports present integrated and scaled data from watersheds to lakes to Great Lakes basinwide.
- Goal 3: Increase outreach and education for the Great Lakes, and provide ongoing K-12 education for students to understand the

benefits and ecosystem functions of the Great Lakes so they are able to make decisions to ensure that restoration investments are enhanced over time.

- Goal 4: Expand the range of opportunities for Great Lakes stakeholders and citizens to provide input to the governments and participate in Great Lakes issues and concerns.
- Goal 5: Work under the goals and objectives of the Great Lakes Water Quality Agreement is coordinated between the United States and Canada through Lakewide Management Plans and other binational processes.

Objectives

- By 2011, opportunities for collaboration, planning, data accessibility and accountability will be increased through the expanded use of Internet-based technology.
- By 2011, an Accountability System will be developed and implemented for the Initiative. The system will integrate and make transparent strategic planning, budgeting, and results monitoring.
- By 2011, a satellite remote sensing program will be implemented to assess Great Lakes productivity and biological (e.g., algal bloom) events.
- By 2011, outreach and education efforts are increased, including identifying and revising existing curricula to incorporate sustainable education needs for the Great Lakes that meet state and other relevant learning standards.
- By 2011, a refined suite of science-based indicators for development of a comprehensive assessment of Great Lakes ecosystem health will be identified, monitoring programs for those indicators will begin to be implemented, and restoration and protection actions tied to those assessments and programs assured.
- By 2011, social media access opportunities for basinwide public involvement in the Initiative will be in place.
- By 2012, education efforts under existing curricula that meet state and other relevant learning standards will be coordinated across states, and a system for tracking student and teacher outreach (quantitatively and qualitatively) for their use.
- By 2012, improved coordination with Canada will take place for programs under the Great Lakes Water Quality Agreement, particularly under the Lakewide Management Plans, which will result

in the achievement of 5-10 priority Lakewide Management Plans goals and actions.

- By 2014, a statistically valid and comprehensive assessment, using a probability-based design, of Great Lakes water resources, will be established. The system will integrate shipboard monitoring, remote sensing, automated sampling, and other monitoring or observing efforts. By 2016, the system will be in place for all of the Great Lakes and it will be capable of providing a scientifically justifiable assessment of Great Lakes water resources.
- By 2014, timely data and information will be provided to decision makers at multiple scales within a framework of established baselines, targets, indicators of progress, and monitoring.

Measures of Progress

- Improvement in the overall aquatic ecosystem health of the Great Lakes using the Great Lakes 40-point scale.[12]
- Number of priority Lakewide Management Plans projects that are completed.
- Number of educational institutions incorporating new or existing Great Lakes protection and stewardship criteria into their broader environment education curricula.[13]

End Notes

[1] Lynn Vaccaro and Jennifer Read, *Vital to Our Nation's Economy: Great Lakes Jobs 2011 Report* (Michigan Sea Grant, 2011). Michigan Sea Grant funds scientific research, education, and extension projects designed to foster science-based decisions about the use and conservation of Great Lakes resources. Michigan Sea Grant also provides access to science-based information about Michigan's coasts and the Great Lakes.

[2] The term Asian carp refers collectively to four species of carp—including bighead and silver carp—that are native to Asia and were first introduced into the United States in 1963. Their rapid expansion and population increase can decrease populations of native aquatic species, in part by consuming vast areas of aquatic plants that are important as food and spawning and nursery habitats.

[3] Pub. L. No. 111-88, 123 Stat. 2904, 2938 (2009). For additional information on the Great Lakes Initiative, see GAO, *Great Lakes Initiative: EPA Needs to Better Ensure the Complete and Consistent Implementation of Water Quality Standards,* GAO-05-829 (Washington, D.C.: July 27, 2005).

[4] H. R. Rep. No. 111-316, at 111 (2009).

[5] The amounts appropriated are approximate because of across the board rescissions and other adjustments.

[6] Executive Order 13340, Establishment of Great Lakes Interagency Task Force and Promotion of a Regional Collaboration of National Significance for the Great Lakes, 69 Fed. Reg. 29043 (May 20, 2004).

[7] The subagencies within the Task Force agencies that are responsible for implementing the GLRI are: the Department of Agriculture's Animal and Plant Health Inspection Service, Forest Service, and Natural Resources Conservation Service; the Department of Commerce's National Oceanic and Atmospheric Administration; the Department of Defense's U.S. Army Corps of Engineers; the Department of Health and Human Services' Agency for Toxic Substances and Disease Registry; the Department of Homeland Security's Coast Guard; the Department of the Interior's Bureau of Indian Affairs, Fish and Wildlife Service, National Park Service, and U.S. Geological Survey; and the Department of Transportation's Federal Highway Administration and Maritime Administration. We refer to all 16 Task Force agencies and subagencies collectively as Task Force agencies in this report, as is the case on the GLRI website http://greatlakesrestoration.us/priorities.html.

[8] Great Lakes Interagency Task Force, *Great Lakes Restoration Initiative Action Plan Fiscal Years 2010-2014*, 09-P-0231 (Washington, D.C.: Feb. 21, 2010.)

[9] GAO, *Great Lakes: EPA Needs to Define Organizational Responsibilities Better for Effective Oversight and Cleanup of Contaminated Areas*, GAO-02-563 (Washington, D.C.: May 17, 2002). No area of concern had had its designation removed—that is, been delisted—by May 17, 2002. EPA has since delisted two areas of concern, one before the creation of the GLRI, in 2006, and one after the creation of the GLRI, in 2013.

[10] GAO, *Great Lakes: Organizational Leadership and Restoration Goals Need to Be Better Defined for Monitoring Restoration Progress*, GAO-04-1024 (Washington, D.C.: Sept. 28, 2004).

[11] EPA Office of the Inspector General, *EPA Needs a Cohesive Plan to Clean Up the Great Lakes Areas of Concern* (Washington, D.C.: Sept. 14, 2009).

[12] We did not include the Department of Housing and Urban Development because officials from that agency and EPA told us that it has not been involved with the Task Force or the GLRI.

[13] According to EPA, phosphorus is a nutrient that controls the amount of algae that will grow suspended in water, and increases in phosphorus can result in increases in algae, which can be detrimental to aquatic life by reducing the amount of available oxygen and sunlight, among other things.

[14] For a complete list of area of concern conditions, click http://www.ijc.org/rel /boards/annex2/buis.htm.

[15] The aquatic nearshore can be considered to begin at the shoreline and extend offshore to the depth at which the warm surface waters typically reach the bottom in early fall, generally 20 to 30 meters deep (i.e., 65 feet 7.4 inches to 98 feet 5.1 inches). Terrestrial nearshore areas range from narrow beaches to inland features affected by lake water.

[16] Pub. L. No. 107-303, 116 Stat. 2355 (2002). Congress consolidated funding for the Great Lakes Legacy Act under the GLRI in 2009. *See* H. R. Rep. No. 111-316, at 110 (2009).

[17] Great Lakes Regional Collaboration, *Great Lakes Regional Collaboration Strategy to Restore and Protect the Great Lakes* (December 2005).

[18] Austin et al., *Healthy Waters, Strong Economy: The Benefits of Restoring the Great Lakes Ecosystem* (Washington, D.C.: The Brookings Institution, September 2007).

[19] H. R. Rep. No. 111-316, at 110-11 (2009).

[20] H. R. Rep. No. 111-316, at 111 (2009).

[21] A 1977 law required EPA to establish a Science Advisory Board to provide scientific advice on a variety of matters. The board, which is a federal advisory committee, has established several standing committees, including the Ecological Processes and Effects Committee.

[22] EPA Science Advisory Board, *Review of Great Lakes Restoration Initiative Action Plan* (Washington, D.C.: Jan. 24, 2012).

[23] For the purpose of this report, we are counting only those projects that were identified in GLAS in May 2013. It is the case that some projects identified in GLAS comprise multiple efforts that are reported as a greater number of separate projects at the GLRI website, http://glri.us/projects/index.html. According to an EPA official, some Task Force agencies may report these individual efforts on the website to provide more comprehensive information on GLRI activities to the public. In May 2013, there were 1,534 GLRI projects identified on the website.

[24] The piping plover is a small shore bird that uses open sandy beaches for its habitat. The piping plover in the Great Lakes watershed was listed as endangered under the Endangered Species Act in 1985.

[25] The total number of GLRI projects identified by Task Force agency in table 2, 1,483, is less than the number of GLRI projects identified by focus area in table 3, 1,487, because 4 GLRI projects are funded by two federal subagencies that were not identified as Task Force agencies and, therefore, are not included in table 2. These subagencies are the Department of Agriculture's Cooperative State Research, Education, and Extension Service and the Department of Health and Human Service's Federal Occupational Health.

[26] Appropriations represent budget authority to incur obligations and make payments from the U.S. Treasury for specified purposes. For budgeting purposes, an allocation means an authorized delegation by one agency of its authority to obligate budget authority and outlay funds to another agency, and it is made when one or more agencies share the administration of a program for which appropriations are made to only one of the agencies. For funds controls purpose, an allocation is a subdivision of an apportionment, which divides the amount available for obligation in an appropriation. Obligations are definite commitments that create a legal liability of the government for the payment of goods and services ordered or received, or a legal duty on the part of the United States that could mature into a legal liability by virtue of actions on the part of the other party beyond the control of the United States. Outlays are the issuance of checks, disbursement of cash, or electronic transfer of funds made to liquidate a federal obligation. See GAO, *A Glossary of Terms Used in the Federal Budget Process*, GAO-05-734SP (Washington, D.C.: Sept. 1, 2005).

[27] EPA did not request applications for grants in the habitat and wildlife protection and restoration focus area in the agency's April 19, 2012, request for applications, noting that the National Oceanic and Atmospheric Administration and Fish and Wildlife Service would be requesting grants in that area that year.

[28] Pub. L. No. 103-62 (1993), *amended by* GPRA Modernization Act of 2010, Pub. L. No. 111-352 (2011). GPRA sought to improve the effectiveness and accountability of federal programs by requiring federal agencies to set goals for program performance, measure results, and report on annual performance compared with the goals. GPRA as amended requires, among other things, that federal agencies develop strategic plans.

[29] EPA in partnership with the Great Lakes Interagency Task Force, *Great Lakes Restoration Initiative Fiscal Year 2011 Report to Congress and the President* (Washington, D.C.: September 2011).

[30] EPA in partnership with the Great Lakes Interagency Task Force, *Great Lakes Restoration Initiative Fiscal Year 2010 Report to Congress and the President* (Washington, D.C.: March 2011).

[31] The National Research Council is the operating arm of the National Academy of Sciences and National Academy of Engineering. Its mission is to improve government decision making and public policy, increase public understanding, and promote the acquisition and dissemination of knowledge in matters involving science, engineering, technology, and health. For more information about the National Research Council click http://www.nationalacademies.org/nrc/index.html.

[32] National Research Council of the National Academies, Committee on Restoration of Aquatic Ecosystems: Science, Technology, and Public Policy, *Restoration of Aquatic Ecosystems: Science, Technology, and Public Policy* (Washington, D.C.: National Academies Press, 1992).

[33] B. K. Williams, R. C. Szaro, and C. D. Shapiro, *Adaptive Management: The U.S. Department of the Interior Technical Guide,* Adaptive Management Working Group, U.S. Department of the Interior (Washington, D.C.: 2009).

[34] *Phragmites australis*, or common reed, is a perennial grass now common in North American wetlands. Invasive phragmites create tall, dense stands that degrade wetlands and coastal areas by crowding out native plants and animals, blocking shoreline views, and reducing access for swimming, fishing, and hunting.

[35] For information on attributes of successful performance measures, see GAO, *Tax Administration: IRS Needs to Further Refine Its Tax Filing Season Performance Measures*, GAO-03-143 (Washington, D.C.: Nov. 22, 2002).

[36] See, for example, GAO, *Environmental Justice: EPA Needs to Take Additional Actions to Help Ensure Effective Implementation*, GAO-12-77 (Washington, D.C.: Oct. 6, 2011).

[37] Pub. L. No. 111-352, §§ 3-4 (2011). Prior to this amendment, the Government Performance and Results Act of 1993 also required agency performance plans to describe the means to be used to verify and validate measured values of performance. Pub. L. No. 103-62, § 4(b) (1993).

[38] GAO, *Performance Plans: Selected Approaches for Verification and Validation of Agency Performance Information*, GAO-GGD-99-139 (Washington, D.C.: July 30, 1999).

[39] GLRI funding recipients are required to submit semiannual progress reports to EPA. According to EPA officials, they do not use these project reports to assess GLRI progress. This is because the progress reports are for administrative management purposes and not for reporting progress made toward the Action Plan measures.

[40] For information on criteria for determining the extent to which an agency's performance plan provides a clear picture of intended performance, see GAO, *The Results Act: An Evaluator's Guide to Assessing Agency Performance Plans* GAO/GGD-10.1.20 (Washington, D.C.: April 1998).

[41] D. C. Evers, J . G. Wiener, C. T. Driscoll, D. A. Gay, N. Basu, B. A. Monson, K. F. Lambert, H. A. Morrison, J. T. Morgan, K. A. Williams, and A. G. Soehl, *Great Lakes Mercury Connections: The Extent and Effects of Mercury Pollution in the Great Lakes Region*, Biodiversity Research Institute, Report BRI 2011-18 (Gorham, Maine: 2011).

[42] GAO-03-143.

[43] According to the Action Plan, the baseline represents the starting point for the measure, and the universe represents all that could possibly be, for example, protected, restored, or enhanced. For example, one measure in the plan is the amount of contaminated sediment remediated in

the Great Lakes. The baseline is 5.5 million cubic yards of sediment, and the universe is 46 million cubic yards.

[44] GAO-03-143.

[45] U.S. Department of Agriculture, U.S. Department of Commerce, U.S. Department of Defense, U.S. Department of Energy, U.S. Department of the Interior, U.S. Environmental Protection Agency, Tennessee Valley Authority, and U.S. Army Corps of Engineers, *Unified Federal Policy for a Watershed Approach to Federal Land and Resource Management*, notice of final policy, 65 Fed. Reg. 62566 (Oct. 18, 2000).

[46] G. H. Stankey, R. N. Clark, and B. T. Bormann, *Adaptive management of natural resources: theory, concepts, and management institutions*, Gen. Tech. Rep. PNW-GTR654: U.S. Department of Agriculture, Forest Service, Pacific Northwest Research Station (Portland, OR: 2005).

[47] Science Subgroup of the Great Lakes Regional Working Group, *Great Lakes Restoration Initiative Adaptive Science-Based Framework for Great Lakes Restoration* (May 21, 2013).

[48] EPA, *Watershed Analysis and Management (WAM) Guide for States and Communities*.

[49] For example, see Williams and Brown, *Adaptive Management;* Stankey et al., *Adaptive Management of Natural Resources;* and RECOVER 2010, *Comprehensive Everglades Restoration Plan Adaptive Management Integration Guide.*

[50] Williams and Brown, *Adaptive Management.*

[51] Williams and Brown, *Adaptive Management.* Williams et al., *Adaptive Management.* Atkinson et al., *Designing Monitoring Programs.* RECOVER 2010, *Comprehensive Everglades Restoration Plan Adaptive Management Integration Guide.* Stankey et al., *Adaptive Management of Natural Resources.*

[52] For example, it states that progress reports should (1) attempt to consider the influence of potentially confounding developments such as natural environmental variation, along with regional and global trends such as climate change, shifts in human population numbers, and changing patterns of land use and (2) include relevant evaluations of the uncertainty and assumptions associated with particular restoration actions and long-term sustainability of ecosystem health.

[53] The fiscal year 2010 report includes a list of five to nine projects in each of the Action Plan focus areas, some of which include a description of progress made.

5 [4]Because some respondents provided examples of more than one type of benefit, the sum of the percent of respondents that provided examples exceeds the 87 percent response rate.

[55] *E. coli* are a large and diverse group of bacteria found in the environment, foods, and intestines of people and animals. Some strains can cause pneumonia and other illnesses in humans who may come into contact with *E. coli* in rivers and lakes contaminated by human or animal feces.

[56] GAO-04-1024.

[57] United Nations Environmental Program and Global Programme of Action for the Protection of the Marine Environment from Land-based Activities, *Ecosystem-based Management: Markers for Assessing Progress* (The Hague, Netherlands: 2006).

[58] The eight indicators are coastal wetlands, phosphorus concentrations, area of concern sediment contamination, benthic health, fish tissue contamination, beach closures, drinking water quality, and air toxics deposition. The ratings for each indicator are added together to create the index score. Improvements in the index would indicate that fewer toxics are entering the food chain; ecosystem and human health is better protected; fish are safer to eat; water is safer to drink; and beaches are safer for swimming.

[59] According to the EPA, the success of an environmental project depends on the quality of the environmental data collected and used in decision making. As part of the process of ensuring that projects have quality data, a grantee creates a document that is intended to serve as a tool for project managers and planners to document the type and quality of data needed for environmental decisions and to describe the methods for collecting and assessing those data. There are neither time limits associated with this process for the GLRI, nor is there a generally accepted length of time for completing these requirements.

[60] The timing of award notification is the timing of when an organization received notice that its GLRI project would be funded.

[61] Because some respondents reported that weather events affected their GLRI projects in more than one way, the sum of the percent of respondents that reported specific effects of weather events exceeds the 26 percent response rate.

[62] For information on atmospheric deposition, see GAO, *Water Quality: EPA Faces Challenges in Addressing Damage Caused by Airborne Pollutants,* GAO-13-39 (Washington, D.C.: Jan. 24, 2013).

[63] The American Society of Civil Engineers represents more than 140,000 members of the civil engineering profession worldwide.

[64] Healing Our Waters – Great Lakes Coalition, *Turning the Tide: Investing in Wastewater Infrastructure to Create Jobs and Solve the Sewage Crisis in the Great Lakes* (Ann Arbor, MI: August 2010). According to the report, data for this statement was provided by the five cities and EPA.

[65] EPA, *Clean Watershed Needs Survey 2008 Report to Congress*, EPA-832-R-10-002 (Washington, D.C.: 2008).

[66] The U.S. Global Change Research Program coordinates and integrates the activities of 13 federal agencies that conduct research on changes in the global environment and their implications for society. For more information about the U.S. Global Change Research Program click http://www.globalchange.gov/.

[67] U.S. Global Change Research Program, Great Lakes Regional Assessment Group, *Preparing for Climate Change, The Potential Consequences of Climate Variability and Change* (Ann Arbor, MI: October 2000). The range of $75 million to $125 million was calculated based on average dredging costs at the time of at least $10 to $12 per cubic yard—about $14 to $16 per cubic yard in 2013 dollars—in a situation with heightened demand for dredging services and 7.5 million cubic yards to 12.5 million cubic removed annually.

[68] U.S. Global Change Research Program, *Global Climate Change Impacts in the United States* (New York, NY: 2009).

[69] The National Academies, Committee on Increasing National Resilience to Hazards and Disasters; Committee on Science, Engineering, and Public Policy; *Disaster Resilience: A National Imperative* (Washington, D.C.: 2012).

[70] For information about climate change adaptation efforts in the Great Lakes region, see R. M. Gregg, K. M. Feifel, J. M. Kershner, and J. L. Hitt, *The State of Climate Change Adaptation in the Great Lakes Region* (Bainbridge Island, WA: EcoAdapt, 2012), accessed June 11, 2013, http://ecoadapt.org/programs/state-of-adaptation/great-lakes-region.

[71] City of Chicago, *Chicago Climate Action Plan* (Chicago, IL: Sept. 19, 2008), accessed June 20, 2013, http://www.chicagoclimateaction.org/.

[72] City of Chicago, *Chicago Climate Change Action Plan Progress Report: the First Two Years* (Chicago, IL: June 23, 2010), accessed June 20, 2013, http://www.chicagoclimateaction.org /pages/ccap_progress_report/81.php.

[73] GAO, *Climate Change: Future Federal Adaptation Efforts Could Better Support Local Infrastructure Decision Makers*, GAO-13-242 (Washington, D.C.: Apr. 12, 2013).

[74] One long-term goal and one objective in the Action Plan address the key factor of insufficient coordination between the United States and Canada. The goal is that work under the goals and objectives of the Great Lakes Water Quality Agreement is coordinated between the United States and Canada through Lakewide Management Plans and other binational processes. The objective is that, by 2012, improved coordination with Canada will take place for programs under the Great Lakes Water Quality Agreement, particularly under the Lakewide Management Plans, which will result in the achievement of five to ten priority Lakewide Management Plans goals and actions.

[75] EPA, *Clean Water State Revolving Fund Programs 2009 Annual Report*, EPA-832-R-10- 001 (Washington, D.C.: June 2010).

[76] GAO, *Water Infrastructure: Approaches and Issues for Financing Drinking Water and Wastewater Infrastructure*, GAO-13-451T (Washington, D.C.: Mar. 13, 2013).

End Notes for Appendix I

[1] The Task Force is chaired by the EPA Administrator, and includes senior officials from the U.S. Departments of Agriculture, Commerce, Defense, Health and Human Services, Homeland Security, Housing and Urban Development, the Interior, State, and Transportation and the White House Council on Environmental Quality. The subagencies within the Task Force agencies that are responsible for implementing the GLRI are: the Department of Agriculture's Animal and Plant Health Inspection Service, Forest Service, and Natural Resources Conservation Service; the Department of Commerce's National Oceanic and Atmospheric Administration; the Department of Defense's U.S. Army Corps of Engineers; the Department of Health and Human Services' Agency for Toxic Substances and Disease Registry; the Department of Homeland Security's Coast Guard; the Department of the Interior's Bureau of Indian Affairs, Fish and Wildlife Service, National Park Service, and U.S. Geological Survey; and the Department of Transportation's Federal Highway Administration and Maritime Administration. We refer to all 16 Task Force agencies and subagencies collectively as Task Force agencies in this report, as is the case on the GLRI website http://greatlakesrestoration.us/priorities.html.

[2] Great Lakes Interagency Task Force, *Great Lakes Restoration Initiative Action Plan Fiscal Years 2010-2014* (Washington, D.C.: Feb. 21, 2010).

[3] GLAS is a mechanism for collecting information to monitor GLRI projects and progress and was created by EPA.

[4] The Nature Conservancy is a nonprofit conservation organization that works to protect ecologically important lands and water for nature and people, and the Great Lakes Indian Fish and Wildlife Commission provides natural resource management expertise and conservation enforcement, among other things, in support of the 11 Great Lakes tribes it represents.

[5] The Healing Our Waters – Great Lakes Coalition consists of more than 120 environmental, conservation, and outdoor recreation organizations, zoos, aquariums and museums, and works to secure a sustainable Great Lakes restoration plan and implement it.

[6] We reviewed GLAS, and not the other sources of progress information—such as from other federal agencies, states, and universities—because EPA officials told us that it is the only

new database created by EPA for the purpose of assessing GLRI progress. We did not use information in GLAS that was entered by users for the purpose of describing progress made by GLRI projects because of concerns raised by EPA about the quality of that information.

[7] EPA Science Advisory Board, *Review of Great Lakes Restoration Initiative Action Plan* (Washington, D.C.: Jan. 24, 2012).

[8] For EPA's list of Great Lakes interested parties, click http://www.epa.gov/greatlakes /parties/states.html.

[9] We conducted interviews with 19 randomly selected, nonfederal individuals at the conference. These individuals were either representatives of nonfederal GLRI stakeholders or Great Lakes interest groups. The number of conference attendees we were able to interview was determined by the number of attendees who were available at the times that we sought out interviewees and were in same location of where we sought out interviewees, and the willingness of the attendees we approached to participate in the interview. Twelve of these individuals recommended subject matter experts.

[10] One of the 17 individuals recommended 21 subject matter experts in a focus area.

[11] We did not invite two subject matters experts to participate. This is because one expert had already provided a lot of information to this report, and we wanted to minimize the response burden. In addition, another expert had demonstrated in a previous interview that he shared the same opinions as a subject matter expert in the same focus area who works for the same organization and who we had already interviewed, and we wanted to minimize the likelihood of duplicate responses. In each case, we interviewed four or more other experts in the relevant focus areas.

[12] EPA in partnership with the Great Lakes Interagency Task Force, *Great Lakes Restoration Initiative Fiscal Year 2010 Report to Congress and the President*, (Washington, D.C.: March 2011), and EPA in partnership with the Great Lakes Interagency Task Force, Great Lakes Restoration Initiative Fiscal Year 2011 Report to Congress and the President (Washington, D.C.: September 2011).

End Notes for Appendix II

[1] The Great Lakes Accountability System is a mechanism for collecting information to monitor GLRI projects and progress and was created by EPA.

End Notes for Appendix III

[1] This is an existing measure [used in the agency's performance plan and performance report, which are required by the Government Performance Results Act, as amended].

[2] These projects represent on-the-ground actions that are being implemented in order to remove beneficial use impairments. For example, sediment removals, Superfund cleanups, habitat projects and others.

[3] This is an existing measure [used in the agency's performance plan and performance report, which are required by the Government Performance Results Act, as amended].

[4] The annual decline is 5 percent per year. This is based on an existing measure [used in the agency's performance plan and performance report, which are required by the Government

Performance Results Act, as amended]. In fiscal year 2010, 2008 data are compared to 2000 data; in fiscal year 2011, 2009 data are compared to 2000 data; and so forth. PCBs are one indicator for a broader suite of persistent toxic substances and one of a number being tracked.

[5] Beaches that the states identify as most frequently used and/or that have the highest risk. In 2008, there were 356 high-priority beaches out a total of 1,411 total beaches in the U.S. Great Lakes.

[6] [Measured in] metric tons per year. Total phosphorus will also be measured. Targeted watersheds will receive focused efforts toward restoration activities, for example, agricultural best management practices.

[7] Biological responses to nutrients loadings are also dependent on other factors such as water temperature, timing and intensity of precipitation, and hydrologic features. Year-to-year variability in these features may mask local improvements in nutrients management. Satellite imagery may provide data for days during which harmful algal blooms are reported by shoreline observers or boaters.

[8] This metric will be added to national surveys for beach managers for 2010. Nuisance algae can include *Cladophora*, harmful algal blooms or other species, all of which are believed to be aggravated by elevated levels of phosphorous in the water.

[9] U.S. Army Corps of Engineers dredges the federal shipping channel at Toledo Harbor each year. This area receives the highest rate of sedimentation in the Great Lakes, coming from the Maumee River watershed. Even small improvements in the rate of sedimentation here would reflect considerable efforts in the watershed to reduce erosion and farm runoff. Alternately, the Corps conducts bathymetric surveys of commercial harbors each year, from which the volume of new fluvial sediment can be calculated for targeted watersheds. Because the Corps does not dredge every location of every harbor each year, the estimated accumulation from a designated area over time will reflect the relative amount of sediments deposited from the tributary. This approach is currently in development.

[10] This measure reflects annual (not cumulative) implementation of conservation practices (from the Environmental Quality Incentives Program and Conservation Technical Assistance program) that will continue to contribute to long term improvements of the listed outcomes.

[11] Also captured under the second measure in focus area one.

[12] This is an existing measure [used in the agency's performance plan and performance report, which are required by the Government Performance Results Act, as amended]. The Great Lakes Index uses select Great Lakes ecosystem indicators (i.e., coastal wetlands, phosphorus concentrations, area of concern sediment contamination, benthic health, fish tissue contamination, beach closures, drinking water quality, and air toxics deposition) and is based on a 1 to 5 rating system for each indicator, where 1 is poor and 5 is good. Improvements in the index and measures would indicate that fewer toxics are entering the food chain; ecosystem and human health is better protected; fish are safer to eat; water is safer to drink; and beaches are safer for swimming.

[13] Educational institutions include: state departments of education, primary and secondary school districts, colleges, universities, zoos, aquaria, museums, and nature/resource centers. Curricula will meet relevant official standards.

In: The Great Lakes Restoration Initiative
Editor: Óskar Strobel
ISBN: 978-1-63117-525-1

Chapter 3

GREAT LAKES RESTORATION INITIATIVE ACTION PLAN FISCAL YEARS 2010-2014*

Great Lakes Interagency Task Force

---

* This is an edited, reformatted and augmented version of a document issued February 21, 2010.

White House Council on Environmental Quality
U.S. Department ofAgriculture
U.S. Department of Commerce
U.S. Department of Health and Human Services
U.S. Department of Homeland Security
U.S. Department of Housing and Urban Development
U.S. Department of State
U.S. Department of the Army
U.S. Department of the Interior
U.S. Department of Transportation
U.S. Environmental Protection Agency

EXECUTIVE SUMMARY

The Great Lakes states and the Great Lakes Basin boundary.

The lands and waters of the Great Lakes are like no other place. In a world where fresh surface water is increasingly in demand, the region contains some 20 percent of it. At a time when people are not looking as much to faraway places for respite, the Great Lakes offer some of the most majestic natural shorescapes on the planet to accommodate them. As a result, people are reconnecting to their beaches, wide-open waters, petro glyphed bluffs, dune ranges and tumbling tributaries like never before. Likewise, these same resources have served as the raw material to build some of the Earth's most legendary cities, create jobs to support families, and contribute to the largest economy in the history of the world.

Still, our expectation that the Great Lakes will continue to meet these needs has resulted in lost flora, fauna, soil, and air and water quality to the point where the ecosystem is showing signs of severe stress and its ability to keep up with these demands is in doubt.

While in the past we have worked to minimize harm, public demand for a new standard of care is surging. That standard of care is that we must leave the Great Lakes better for the next generation than the condition in which we inherited them. We must continue to go beyond minimizing harm to proactively rehabilitating the Great Lakes. Only then will they be able to keep providing jobs, recreation and sanctuary.

Understanding this, U.S. President Barack Obama and U.S. Environmental Protection Agency (EPA) Administrator Lisa Jackson, in collaboration with 15 other federal agencies, have made restoring the Great Lakes a national priority. Signaling a commitment beyond measure of past promises, in February 2009, President Obama proposed $475 million for a Great Lakes Restoration Initiative (Initiative). This Action Plan describes how the Initiative will be executed from 2010 through 2014.

The Initiative is not intended to be another grand statement about the Great Lakes; it is intended to operationalize those statements. It builds on countless hours by elected, agency, business, public interest and other leaders, which resulted in the Great Lakes Regional Collaboration Strategy (GLRC Strategy). The GLRC Strategy provides a framework for the Action Plan, and the Action Plan is just that: an action driver. It articulates the most significant ecosystem problems and efforts to address them in five major focus areas:

- **Toxic Substances and Areas of Concern,** including pollution prevention and cleanup of the most polluted areas in the Great Lakes (see pages 19-21 for Measures of Progress with specific, quantifiable targets; and Principal Actions)

- **Invasive Species,** including efforts to institute a "zero tolerance policy" toward new invasions, including the establishment of self-sustaining populations of invasive species, such as Asian Carp (pages 24-26)
- **Nearshore Health and Nonpoint Source Pollution,** including a targeted geographic focus on high priority watersheds and reducing polluted runoff from urban, suburban and, agricultural sources (pages 29-30)
- **Habitat and Wildlife Protection and Restoration**, including bringing wetlands and other habitat back to life, and the first-ever comprehensive assessment of the entire 530,000 acres of Great Lakes coastal wetlands for the purpose of strategically targeting restoration and protection efforts in a science-based manner (pages 33-35)
- **Accountability, Education, Monitoring, Evaluation, Communication and Partnerships,** including the implementation of goal- and results-based accountability measures, learning initiatives, outreach and strategic partnerships (pages 38-39)

The Action Plan identifies goals, objectives, measurable ecological targets, and specific actions for each of the five focus areas identified above. The Action Plan will be used by federal agencies in the development of the federal budget for Great Lakes restoration in fiscal years 2011 and beyond. As such, it will serve as guidance for collaborative restoration work with participants to advance restoration. The Action Plan will also help advance the Great Lakes Water Quality Agreement with Canada.

The Initiative is not the only tool in the toolbox, however. Traditional infrastructure financing under Clean and Drinking Water State Revolving Funds, and Superfund cleanup enforcement, for example, represent work outside the Initiative's scope. This work, however, continues to be essential to Great Lakes protection and restoration and EPA is working with states and tribes to ensure that these high priority activities are properly targeted whenever possible to help further clean-up of the Great Lakes.

This plan also aims to build upon the significant amount of work already accomplished by states and other partners in protecting and restoring the Great Lakes. In addition, it tries to minimize the duplication of effort by focusing on high priority work that has already been identified in the many programs and strategies already in place around the Great Lakes Basin. This includes work under the Great Lakes Regional Collaboration Strategy, individual states' Great Lakes restoration plans, and many others.

Under the Initiative, EPA will administer funding individually and with other federal agencies. Funds will be used to implement priority federal projects as well as other programs undertaken by nonfederal entities that support the Action Plan. Funding will be provided through grants and cooperative agreements or through interagency agreements that allow the transfer of funds to other federal agencies for subsequent use and distribution. Most grants will be issued competitively. Annual reports to the President, beginning in 2011, will describe accomplishments to date, action planned for the upcoming year, and progress toward meeting ecosystem goals and targets. These efforts will underpin the Action Plan's operating principles:

1) **Accountability:** The Initiative is an unprecedented opportunity to heal the ecosystem. With this unprecedented opportunity comes unprecedented responsibility, however, for all of us to demonstrate we are achieving the results intended in the Action Plan. We will use transparent means of demonstrating how public dollars are being invested as directed by the best available science.
2) **Action:** This distinguishes the Initiative from so many other efforts in the past. For the most part, we know the problems facing the Great Lakes. The Initiative supports our resolve to solve them. While there is a place for monitoring, sampling, surveying and planning under the Initiative, these activities must have a bias for action, that is, real on-the-ground and in-the-water project work that will breathe new life into the ecosystem. Projects ready to go with existing plans will receive preference over efforts that require extensive new planning.
3) **Urgency:** Study after study shows the health of the lakes is in jeopardy. While we have been working heroically to hold the line over the years to prevent their further decline, we have been entrusted with powerful new resources to rehabilitate the lakes: leadership at the highest levels demanding action, policy tools, and the promise of significant financial resources. With these, there are no more reasons for delay. After all, the Great Lakes have endured some 150 years of abuse. Though it will take time for the ecosystem to respond to our actions today, there is not another minute to lose in restoring these magnificent waterways for tomorrow. To the extent that actions undertaken now increase the resiliency of the ecosystem, they will also help the Great Lakes adapt to climate change.

This Action Plan — including the funding and other policy tools that will be used to leverage its outcomes — is intended to realize our goals: that the fish are safe to eat; the water is safe to drink; the beaches and waters are safe for swimming, surfing, boating and recreating; native species and habitats are protected and thriving; no community suffers disproportionately from the impacts of pollution; and the Great Lakes are a healthy place for people and wildlife to live.

A National Treasure

The Great Lakes — the largest group of freshwater lakes on Earth — are true wonders of the world. An important part of the physical landscape and cultural heritage of North America, the Great Lakes hold 95 percent of the United States' surface fresh water. Shared with Canada, these "freshwater seas" boast more than 10,000 miles of magnificent coastline and 30,000 islands and provide drinking water, transportation, power and a wide array of recreational opportunities. The region's four-season climate, uniquely influenced by the Great Lakes, supports boating, fishing, diving, beach enjoyment and other forms of recreation that support the region's proud outdoor heritage.

The environment of the Great Lakes region is blessed with wide swaths of forest and wilderness areas, rich agricultural land, hundreds of tributaries, thousands of smaller lakes, and extensive mineral deposits. The region's sand dunes, coastal marshes, rocky shorelines, lakeplain prairies, savannas, forests, fens, wetlands and other landscapes contain features that are globally unique or best represented within the Great Lakes basin. For example, the world's largest freshwater dunes line the shores of Lake Michigan.

The region's glacial history and the influence of the lakes themselves create unique conditions that support a wealth of biological diversity, including over 200 globally rare plants and animals and more than 40 species that are found nowhere else in the world. Rare species making their home in the Great Lakes region include the world's last known population of the white catspaw pearly mussel, the copper redhorse fish and the Kirtland's warbler. The Great Lakes environment supports a world-class fishery, with an estimated 180 species of native fish, including small- and large-mouth bass, muskellunge, northern pike, lake herring, whitefish, walleye and lake trout.

If the Great Lakes region were its own nation, it would house the largest economy second only to the United States itself, providing transportation for raw materials and finished goods; fresh water for our industries; drinkin water for our communities; and recreation for the basin's more than 30 million citizens. The 4.3 million recreational boats registered in the eight Great Lakes states generate nearly $16 billion in spending on boats and boating activities in a single year. That spending directly supports 107,000 jobs, a figure that grows to nearly 250,000 when secondary impacts are taken into consideration.

The Great Lakes Restoration Challenge

The Great Lakes impact our way of life, as well as all aspects of the natural environment, from weather and climate to wildlife and habitat. Yet for all their size and power, the Great Lakes are not as resilient as they look. In the past, their fragile nature was not recognized and the lakes were mistreated for economic gain, placing the ecosystem under tremendous stress from our activities. Today, we understand that our health and our children's future depend on our collective efforts to wisely manage this complex ecosystem.

History has shown us that the Great Lakes are highly sensitive to biological and chemical stresses. While restoration progress has been made through years of concerted effort and expenditures on the part of federal, state, tribal and local governments and other stakeholders, that progress is slowing or even reversing. The Great Lakes face a number of serious challenges. The most significant of these include toxic substances, invasive species, nonpoint source pollution and nearshore impacts, habitat and species loss, and a need for better information to guide decision making.

Although releases of toxic pollutants have been reduced significantly over the last 30 years, there is a legacy of contamination in sediments and continuing inputs through rivers and air. Excessive levels of contaminants are still found in fish throughout the system. As a result, all Great Lakes states and the province of Ontario issue annual fish consumption advisories. Mercury and other pollutants continue to enter the Great Lakes from nearby and global sources through air deposition. Newly recognized chemicals of concern are also being identified as potential threats to the chemical integrity of the Great Lakes. Thirty toxic hotspot "Areas of Concern" (AOCs) in the United States are still in need of cleanup after more than 20 years; only the Oswego River has been removed from the AOC list.

Aquatic and terrestrial invasive species continue to cause ecological and economic damage, and greatly complicate efforts to restore the Great Lakes. New aquatic species of invaders have arrived at the rate of about one every eight months, adding to the more than 180 already established in the basin. In recent years this rate has slowed but the door to new arrivals has not yet been shut. Enhanced prevention and control efforts are necessary to stop new invasive species like Asian Carp from becoming established in the Great Lakes.

Pollution from nonpoint sources contributes to impaired water quality and excess nutrients. Many of our coastal areas also suffer from sewer overflows that contaminate the water and close the beaches.

Habitat destruction and degradation due to development, competition from invasive species, alteration of natural lake level fluctuations and flow regimes, poor coastal development planning and land management, and habitat fragmentation have negatively impacted wildlife. This has led to altered food webs, a loss of biodiversity, and poorly functioning ecosystems. Yet, opportunities for the protection and restoration of critical habitat exist throughout the basin.

While the Great Lakes region has been a leader for innovative science and advances in natural resource management, there are still significant gaps in knowledge about ecological processes and key indicators of ecosystem health. Efforts must be strategically chosen in order to obtain the additional information needed to inform implementation activities, assist tracking and reporting of progress, and to identify adaptive management actions. The Great Lakes also face new and emerging problems such as the effects of climate change, including potentially changing long-term Great Lakes water levels and the timing and duration of ice cover.

Collectively, these problems have seriously compromised the environmental health of the Great Lakes. As a result, there is a new sense of urgency for action to address the highest priorities for restoring and protecting the Great Lakes. This document outlines a strategic multi-year approach to address those urgent problems.

WHAT IS "RESTORATION"?

Restoration is the process of assisting the recovery of an ecosystem that has been degraded, damaged or destroyed. For purposes of this Initiative, restoration includes ecosystem protection, enhancement, rehabilitation and remediation. A restored ecosystem is resilient; its chemical, physical, and biological functions and processes provide the requisite conditions for life. A restored ecosystem contains sustainable populations of native plant and animal species and their habitats. Potential threats or further damage have been eliminated or reduced as much as possible and the restored ecosystem is able to withstand future threats. Restoration *is not* the attempt to change the ecosystem to pre-European settlement conditions; however, a restored ecosystem does attempt to emulate those conditions to the extent possible under present-day chemical, physical and biological conditions.

Restoration of degraded, damaged or destroyed water and lands is more costly than protection of resources before damage occurs. Therefore, this Initiative recognizes the wisdom of supporting ecological protection. **Protection is defined as actions taken to prevent stress to ecosystems**. Actions include the establishment of stewardship partnerships to manage and monitor habitats and species; construction of physical barriers to prevent damage to sensitive areas; implementing best management practices to protect ecosystems; and, acquisition of land and protection agreements such as easements. The Initiative acknowledges that land acquisitions and protection agreements alone will not protect ecosystems from threats such as invasive species, encroaching development, or airborne toxic pollutants. Acquisition and protection agreements must be accompanied by appropriate stewardship and management actions.

TAKING ON THE CHALLENGE: THE GREAT LAKES RESTORATION INITIATIVE

When running for president, Barack Obama issued a campaign promise to protect and restore the Great Lakes. The pledge built upon a May 2004 Executive Order that created the Great Lakes Interagency Task Force to coordinate federal restoration efforts. In 2005, some 1,500 stakeholders created the Great Lakes Regional Collaboration Strategy (GLRC Strategy) that outlines challenges facing the lakes, a framework for restoration and protection, and a common set of recommended solutions across eight priority issue areas:

- Aquatic Invasive Species
- Habitat/Species
- Coastal Health
- Areas of Concern (AOCs)/Sediments
- Nonpoint Source

- Toxic Pollutants
- Indicators and Information
- Sustainable Development[1]

Released in May 2009, the President's FY 2010 budget included $475 million for a new Great Lakes Restoration Initiative (Initiative), strategically targeting programs and projects to address the most significant problems in the Great Lakes ecosystem to demonstrate measurable results. EPA, in concert with its federal partners on the Task Force and other stakeholders, is leading the development and implementation of this Initiative and will administer the funding. Building upon the extensive planning and collaboration that was done by the Task Force and a wide variety of stakeholders and non-governmental partners in development of the GLRC Strategy, the Task Force developed a plan for FY 2010.[2]

The Initiative presents an unprecedented opportunity to systematically tackle the environmental challenges that have persisted for decades. In July and August 2009, EPA held a series of public meetings with agencies and stakeholders in the Great Lakes states to get public feedback on the highest priority issues in each focus area, both basinwide and locally, and to get suggestions for how to address these issues to maximize results under the Initiative. This Action Plan incorporates those basinwide and local priorities into a five year time frame for action. Federal agencies began implementing that FY 2010 plan following Congressional appropriation of the President's $475 million request.[3]

THE PATH FORWARD: THE ACTION PLAN

Plan Overview

This Great Lakes Restoration Action Plan (Action Plan) outlines methods and actions to advance implementation of the Initiative through FY 2014 and will help protect and restore the chemical, physical and biological integrity of the Great Lakes Basin ecosystem.

Five principal focus areas have been identified which encompass the most significant environmental problems in the Great Lakes (other than water infrastructure) for which urgent action is required. These include[4]:

- Toxic Substances and Areas of Concern
- Invasive Species
- Nearshore Health and Nonpoint Source Pollution
- Habitat and Wildlife Protection and Restoration
- Accountability, Education, Monitoring, Evaluation, Communication and Partnerships

Each focus area in this Action Plan includes a problem statement, goals, objectives, measures of progress and targets, and the principal actions in support of the objectives. On an annual basis, federal agencies are expected to identify the specific actions they will take to implement the Action Plan. Federal agencies also will minimize overhead in carrying out the Action Plan and, to the extent feasible, will use existing staff. EPA will assure that the Action Plan's goals, objectives, and targets are aligned with those of Great Lakes state, municipal and tribal g overnments.

Within the five focus areas, the Action Plan will address the highest priority projects. It is the intent of the Initiative's federal agencies to target efforts and funds to these projects in a way that maximizes results. Targeted, cooperative efforts are necessary to ensure meaningful progress on many of the complex and costly issues that have plagued the Great Lakes for decades. Some issues exist basinwide (e. g., invasive species, nonpoint source pollution,) and require broad, expansive action, while others are more localized (e. g., Areas of Concern, habitat) and will have site-specific remedies.

In each focus area there are efforts which will be given special attention.

- In the focus area of Toxic Substances and Areas of Concern, efforts will be targeted to remediate contaminated sediments and to address other major pollution sources in order to restore and delist the most polluted sites in the Great Lakes basin.
- In the focus area of Invasive Species, efforts will be targeted to institute a "zero tolerance policy" as a long term goal toward new invasions, including the development of ballast water technology, an early detection surveillance program, and a rapid response capability to address threats from new invasive species such as Asian Carp.
- In the focus area of Nearshore Health and Nonpoint Source Pollution, efforts will be targeted geographically to focus on watersheds of extreme ecological sensitivity (such as the Green Bay/Fox River, Genesee River, Maumee River, St. Louis River, and Saginaw River,

places where environmental problems and their solutions have been clearly identified).

- Efforts will target implementation of lakewide biodiversity conservation blueprints and restoration of important species such as the Lake Sturgeon and the Piping Plover.
- In the focus area of Accountability, Education, Monitoring, Evaluation, Communication and Partnerships, efforts will include implementation priority Lakewide Management Plan projects for restoring the lakes, as well as establishment of quality goals and results-based accountability measures, learning initiatives, outreach and strategic partnerships.

Detailed information about these efforts is included in the focus area descriptions which follow in a subsequent section of this Action Plan.

Building on Past Progress, Integrating with Current Programs

This Action Plan includes many elements of the Strategy and draws upon the ecological priorities, goals and objectives of numerous pre-existing issue or area-specific plans and programs that have been developed by federal, state, tribal, local and non-governmental stakeholders, such as:

- Plans under the Great Lakes Water Quality Agreement, including Lakewide Management Plans and Remedial Action Plans for Areas of Concern
- Comprehensive Management Plan for St. Clair River and Lake St. Clair
- Great Lakes Binational Toxic Strategy
- State Wildlife Action Plans
- U.S. Fish and Wildlife Service (FWS) National Fish Habitat Action Plan
- A Joint Strategic Plan for Management of Great Lakes Fisheries
- Plans under the U.S. - Canada Convention on Great Lakes Fisheries
- Partners in Flight North American Land Conservation Plans
- North American Waterbird Conservation Plan
- North American Waterfowl Management Plan
- U.S. Shorebird Conservation Plan
- Great Lakes Coastal Wetland Monitoring Plan

- Great Lakes Fishery Plan
- Endangered Species Recovery Plans
- Coastal Management Plans under the Coastal Zone Management Act
- Plans under the Coastal and Estuarine Land Conservation Program
- Source Water Protection Plans
- Tribal-Federal Environmental Agreements

This Action Plan takes a step toward integrating and aligning these and other plans, including individual states' Great Lakes restoration plans. Our goal is to minimize the duplication of effort while using the Initiative to do high priority work that has already been identified or that may be incomplete, under other programs. As such, the Action Plan represents our best attempt to capture the broad consensus of the Great Lakes community's collective commitment to advance Great Lakes protection and restoration. This broad consensus is the result of years of planning finally capped by 18 meetings in summer 2009 with representatives of states, tribes, cities and other stakeholders. We expect that these representatives and others will be able to continue providing input to federal agencies for implementation of priority Great Lakes actions by participating in these and other ongoing planning efforts. The Great Lakes Restoration Initiative will also support the National Policy and Implementation Plan being developed in response to the President's memorandum on a "National Policy for the Oceans, Our Coasts, and the Great Lakes."

In addition to new federal funding through GLRI, the Initiative will rely on partnerships to leverage and harness a wider set of resources for the protection and restoration of the Great Lakes. In many instances, the most effective solutions to the challenges facing the Great Lakes will require effective use of non-GLRI baseline federal funding, federal regulatory or other policy tools, and the significant regulatory and policy tools and resources of states, tribes, and other non-federal partners. These efforts, summarized below, are complementary to GLRI efforts.

- **Non-GLRI baseline federal funding.** For example, construction of water infrastructure for treatment and conveyance of drinking water and wastewater.
- **Federal regulatory or other policy tools.** For example, national rulemakings and permitting that reduce the risk of future invasions of ANS; permitting activities under the Clean Water Act; or regulatory means to reduce atmospheric mercury deposition to the Great Lakes.

- **Tools and resources of non-federal partners.** For example, stewardship of properties by a state, tribe or non-governmental organization for the purpose of enhancing habitat protection or connectivity; or implementation of source water protection plans for drinking water treatment facilities with intakes that draw from surface or groundwater within the Great Lakes basin.

Federal Integration and Coordination

The primary means for ensuring that the Action Plan is developed and implemented in an integrated manner is the Great Lakes Interagency Task Force, chaired by EPA.

Created by presidential Executive Order in 2004, the Interagency Task Force (Task Force), through the efforts of the Regional Working Group (RWG), coordinates federal Great Lakes efforts, including some 140 federal programs that address Great Lakes issues. The Task Force includes 11 of the agency and cabinet organizations which manage those programs: EPA, State, Interior, Agriculture, Commerce, Housing and Urban Development, Transportation, Homeland Security, Army, Council on Environmental Quality, and Health and Human Services.

IMPLEMENTING THE ACTION PLAN

Though a national treasure of magnificent grandeur, we know the Great Lakes are still vulnerable to misuse and exploitation. Degradation of the lakes and their supporting ecosystem happened over the course of more than a century. Even with nature's resiliency, discernable widespread improvements will likely also take time. The goals, objectives, measures of progress and other efforts identified in this Action Plan will help ensure quantifiable annual outputs are in place that will lead to long-term restoration outcomes, even if such outcomes will not be immediately measurable on an ecosystem scale.

Tracking Progress and Using the Best Available Science

The Task Force will implement an accountability system to routinely track, measure and report progress pursuant to the Action Plan. Recipients of

Initiative funding will be required to provide routine progress reports, at least semi-annually, on their individual projects as well as progress toward the goals, objectives, and measures of the Initiative. EPA will collect that information and report on overall progress toward attaining the goals and objectives of the Action Plan. EPA will work with the Task Force to identify, adapt and modify activities in future years based on performance and newly identified needs. These will be included in the development of funding plans for subsequent years.

Beginning in March, 2010, EPA will provide annual reports that provide funding allocations by federal agency and that identify any adjustments from annual requests. In addition, starting in 2011, EPA will work with its Task Force partners to provide an Annual Report to the President on accomplishments in achieving the Action Plan's outcomes and measures. The report will also compare funding allocations among participating agencies from fiscal year to fiscal year.

Scientific integrity at all levels is critical to the success of Great Lakes restoration. Therefore, the Initiative will use the best available science to prioritize and implement actions, as well as to track progress. To do this, EPA and the Task Force will take all appropriate steps so that the Action Plan and its programs and projects optimize the likelihood of successful restoration at relevant scales. This will include engaging an independent scientific review panel to review the scientific credibility of the Action Plan so that the best available science guides efforts to restore the Great Lakes. Some of the distinct programs and projects in this Plan may also require a more specific peer review. These scientific reviews will be scaled as necessary to establish the scientific justification and credibility for the Action Plan's goals, objectives, and measures and actions The GLRI will also ensure that the rate of progress is scientifically validated, and scientific research will guide any revisions to restoration priorities as well as enable the Initiative to adapt and modify activities when necessary.

Project Selection

The following criteria and principles will guide selection of pro grams[5] and projects pursuant to this Action Plan. All agencies and prospective grant applicants should consult the respective Requests for Proposals (RFPs) issued each year to solicit grant proposals, as project selection criteria and principles may change from year to year.

- Ability to strategically achieve measurable environmental outcomes linked to the highest priority issues
- Ability to advance applicable ecological priorities of existing plans, such as Lakewide Management Plans, Remedial Action Plans for Areas of Concern, as well as other relevant national and regional coordinated strategic planning efforts, as mentioned above
- Feasibility of prompt implementation, including a bias for projects that are both ready-to-go and will have results soon (however, some funding will be used for planning and design to ensure cost effective implementation and for monitoring, particularly where it is needed to establish baseline conditions and/or to better understand environmental problems to inform implementation actions)
- Observable local impacts, especially for projects at the "field" level
- Strong bias for interagency/inter-organizational coordination and collaboration
- Support new work, or enhance (but do not replace) existing Great Lakes baseline activities
- Public support
- Ability to leverage non-federal resources
- Promotion of long-term societal, economic, and environmental sustainability
- Minimization of transaction costs

Projects and activities must also meet standards for:

- best available science
- experience, ability, and authority of the funding recipient to properly perform the work;
- reasonableness of project costs; and
- measuring progress and success.

The Great Lakes Interagency Task Force used the criteria above to develop the focus areas of the Initiative, evaluate programs and projects, and create provisional funding allocations.

Work within each focus area will be accomplished through federal interagency cooperation, and by working closely with states, tribes, local government, academia, NGOs, and other stakeholders in the Great Lakes basin, as well as Canadian representatives. As this Action Plan encompasses numerous national and regional coordinated strategic planning efforts and their associated plans, participating agencies must assure that they continue to support their existing base program activities in the Great Lakes and that added Initiative support will not be used to supplant their existing base funding. As a starting point for identifying their base funding, agencies have been asked to use the 2009 Office of Management and Budget's Great Lakes Restoration Initiative Crosscut Budget Report to Congress. Final funding allocations are dependent upon actual appropriations.

Funding and Grants Cycle

Through this Action Plan and collaboration among EPA and the other agencies on the Great Lakes Interagency Task Force, and with input from Great Lakes stakeholders, the distribution of funds will be directed to maximize Great Lakes restoration and protection. Resources will be directed to the most significant issues and opportunities.

To be positioned to fund projects through grants[6] as soon as possible after an annual appropriation is made, EPA will collaborate with the Interagency Task Force to do as much up-front work as possible, including issuance of a RFP as soon as possible each fiscal year. The planning process assumes that an appropriation for grant funding for states, tribes, local governments, and other organizations can be available early in each fiscal year.

Most EPA grants will be issued competitively pursuant to RFPs addressing the five focus areas.[7] These expressions of needed work are included in numerous issue-based and geographically focused plans as referenced above. Should significant problems and issues need to be addressed outside of the five focus areas, a competitive grant program would be used to fill gaps, cut across or overlap focus areas, address unanticipated areas, or facilitate innovation. Important criteria for grant selection will include a demonstration of the ability to commence work expeditiously to connect the project to Great Lakes priorities.

Following annual appropriations, EPA will act upon expert recommendations and will select proposals and issue grants for EPA programs. Upon routine implementation of the Initiative, EPA believes that if it were to receive an appropriation by October 1 of a given fiscal year, the first grants could be issued early in the new calendar year, with others issued in the course of the year. One or more additional RFPs would be issued as needed throughout the year.

Several other members of the Interagency Task Force are also expected to select proposals, issue grants, and provide other assistance with funding from the Initiative. Each agency would provide assistance following its own applicable procedures, but would require special Initiative reporting provisions that will be outlined in the Interagency Agreement. EPA would be given the opportunity to review other agencies' Requests for Proposals (RFPs) and proposed project selections. Likewise, EPA would share its RFPs and proposed project selections to avoid duplication. To assist stakeholders in finding assistance opportunities pursuant to the Initiative in a single location, EPA developed and will periodically update the Task Force's Funding Guide. Grant issuing agencies will ensure that appropriate results and accountability information is incorporated into public reports and provided to oversight groups.

FOCUS AREAS

The following describes the five focus areas around which significant ecological stresses and efforts to address those stresses will be organized:

- Toxic Substances and Areas of Concern
- Invasive Species
- Nearshore Health and Nonpoint Source Pollution
- Habitat and Wildlife Protection and Restoration
- Accountability, Education, Monitoring, Evaluation, Communication and Partnerships

Each focus area includes subsections that provide a problem statement to describe stresses, long-term goals[8] for addressing those stresses, objectives, measures of progress to ensure efforts are on track for implementing longterm goals and principal actions to achieve that progress. This section is also helpful to prospective grantees and others who might receive Initiative funds to ensure what they plan to propose is in keeping with overall Initiative efforts.

Some priorities cut across focus areas. For example, the Task Force will "geographically target" activities, such that federal, state and other stakeholders can leverage efforts to restore areas that are highly degraded and of high ecological importance in the Great Lakes. Geographic targeting across the focus areas is expected at places such as the Genesee River, Green

Bay/Fox River, Maumee River, St. Louis River, and Saginaw River watersheds.

Another cross-cutting priority will be environmental justice and work to reduce disproportionate ecological impacts. Environmental justice, as a priority, will be handled under some of the individual focus areas (e. g., cleanups in Areas of Concern, which may correspond to areas of potential environmental justice concern). While we do not anticipate creating a separate funding source for work in this area, special consideration will be given to efforts that address these important priorities.

Similarly, projected impacts of climate change on the Great Lakes have implications across all focus areas. Climate change impacts and the needs of the Great Lakes community to adapt to those impacts will be assessed and addressed by GLRI projects and programs where appropriate. To the extent that actions undertaken as part of this Initiative increase ecosystem resiliency, they will also help the Great Lakes ecosystem adapt to climate change.

Focus Area 1: Toxic Substances and Areas of Concern

Problem Statement

Although many point sources of pollution — discharges from discernible, often endof-pipe conduits — have been reduced, legacy contamination remains. "Legacy contamination" is pollutants largely left over from past practices, but that continue to recirculate through the ecosystem. Such legacy pollutants, often persistent toxic substances (PTS), such as mercury and polychlorinated biphenyls (PCBs), continue to be present at levels above those considered safe for humans and wildlife, warranting fish consumption advisories in the Great Lakes, connecting channels, and Midwestern and New York interior lakes.

Urban communities living in or near these areas and indi genous communities that still live off the land in the basin are particularly at risk from disparate impacts on health from pollution in these areas, and from consuming contaminated fish. Continuing sources of persistent toxic substances include releases from contaminated sediments; industrial and municipal point sources; nonpoint sources including atmospheric deposition, agricultural and urban runoff, and contaminated groundwater; and cycling of the chemicals within the lakes.

Action Illustration:
Electronic Waste and Unwanted Medicines

During the 2008 Earth Week Campaign U.S. EPA held an electronic and unwanted medicines waste collection campaign with the goal of collecting 1 million pounds of e-waste and one million pills of expired and unused medicines in the Great Lakes basin. Approximately $500,000 in grants was awarded to states, tribes, and local governments. Over 100 entities participated in the week-long campaign across the basin. The campaign goals were far exceeded with approximately 5 million pounds of electronic waste and 5 million pills collected. With funding and support from the GLRI, over the next two years, we anticipate broad participation in similar collection events throughout the Great Lakes basin, resulting in at least 20 million pounds of electronic waste and 20 million pills of unwanted medicines collected and kept safely out of the Great Lakes environment. Beyond the collection events, the principal goals in this effort include raising awareness in Great Lakes communities regarding the importance of safe disposal of electronic waste and unwanted medicines, and continuing to cultivate the development of ongoing and sustainable hazardous waste collection programs within the Great Lakes basin, as well as making communities and businesses aware of existing collection programs.

Efforts to restore degraded conditions in the 30 U.S. Great Lakes Areas of Concern (AOCs) are underway using a variety of funding sources including those under the Great Lakes Legacy Act, Superfund and other tools, but much more needs to be done, including the remediation of an estimated 43 million cubic yards of contaminated sediments, which are the main cause of beneficial use impairments in the majority of the AOCs.

In addition to the well-known toxicants like mercury, PCBs, and banned pesticides, there are chemicals of emerging concern that have been detected in the Great Lakes over the past several years, which may pose threats to the ecosystem. Some such chemicals may include flame retardants, surfactants, pharmaceuticals and personal care product constituents.

Beyond protecting human and ecosystem health related to fish consumption, activities in this focus area that address current and new pollution threats can also help protect drinking water sources.

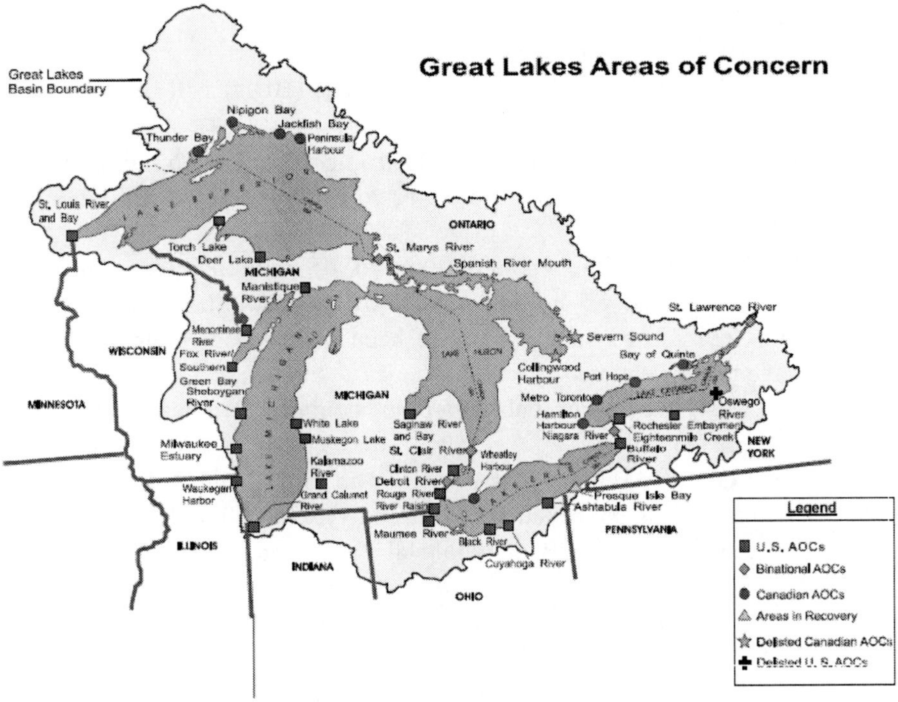

Long Term Goals

- Goal 1: Areas of Concern are cleaned up, restoring the areas and removing the beneficial use impairments.
- Goal 2: The release of toxic substances in toxic amounts is prevented and the release of any or all persistent toxic substances (PTS) to the Great Lakes basin ecosystem is virtually eliminated.
- Goal 3: Exposure to toxic substances from historically contaminated sources is significantly reduced through source reduction and other exposure reduction methods.
- Goal 4: Environmental levels of toxic chemicals are reduced to the point that all restrictions on the consumption of Great Lakes fish can be lifted.
- Goal 5: The health and integrity of wildlife populations and habitat are protected from adverse chemical and biological effects associated with the presence of toxic substances in the Great Lake Basin.

Objectives

- By 2014, delist five Areas of Concern.
- By 2014, 46 Beneficial Use Impairments (BUIs) will be removed in Areas of Concern.
- By 2011, 15 million pounds of electronic waste and 15 million pills of unwanted medicines will be collected or their release will have been prevented.
- By 2014, 45 million pounds e-waste, 45 million pills of unwanted medicines, and 4.5 million pounds of household hazardous waste in the Great Lakes basin will have been collected or their release will have been prevented.
- By 2014, 9.4 million cubic yards of contaminated sediments will be remediated.
- Through 2014, an annual average of up to 5% annual decline will be maintained or improved for the trend (year 2000 and on) in average concentrations of PCBs in whole lake trout and walleye samples.

Action Illustration: Cleaning Up Areas of Concern

Thirty of the originally identified 31 toxic hotspot "Areas of Concern" (AOCs) have been on official lists for cleanup for more than 20 years. In all that time, only the Oswego River AOC has been delisted and only the Presque Isle AOC has attained a "recovery" status. In both cases, local governments implemented pollution control programs and a strong local citizenry developed community-based volunteer programs to assess, monitor and improve environmental conditions. AOC cleanup is complex for several reasons - due to the nature of jurisdictional issues among multiple levels of government, the complexity of the many environmental programs that are being brought to the task, and the immensity of the environmental issues being faced. For example, prior analyses have placed the costs for sediment clean-up alone across all AOCs in the Great Lakes basin on an order of $1 — 4 billion. In addition to sediment cleanups, AOCs face other costly and intractable environmental problems, such as agricultural and urban nonpoint source pollution, combined sewer overflows and other manifestations of aging water infrastructure, wetlands loss, and others. Solutions to these problems often require decades of work, all the time coordinating among multiple partners and authorities to harness and synchronize the significant resources that are necessary over several years.

Measures of Progress

The Great Lakes Restoration Initiative will si gnificantly accelerate pollution prevention and reduction in the Great Lakes ecosystem. The measures by which pro gress will be evaluated in this focus area are:

Measure	Baseline/ Universe[9]	2010 Target	2011 Target	2012 Target	2013 Target	2014 Cumulative Target
Number of Areas of Concern in the Great Lakes where all management actions necessary for delisting have been implemented (cumulative).	Baseline: 1 AOC Universe: 31 AOCs	1 AOC (no change)	1AOC	3AOCs	4 AOCs	5 AOCs
AOC BUIs removed (cumulative).[10]	Baseline: 11 BUIs Universe: 261BUIs	20 BUIs	26 BUIs	31 BUIs	41 BUIs	46 BUIs
BUI delisting project starts at AOCs (cumulative).[11]	Baseline/Univer se: 30 national and binational AOCs Universe: 261BUIs	60 projects	80 projects	110 projects	140 projects	170 projects
Cubic yards (in millions) of contaminated sediment remediated in the Great Lakes (cumulative).[13]	Baseline: 5.5 million cubic yards (2007) Universe: 46 million cubic yards	6.3 million cubic yards	7.0 million cubic yards	7.2 million cubic yards	8.6 million cubic yards	9.4 million cubic yards
Pollution (in pounds) collected through prevention and waste minimization projects in the Great Lakes basin (cumulative).	Baseline: 0	10 million pounds	15 million pounds	25 million pounds	35 million pounds	45 million pounds
Cumulative percentage decline for the long term trend in average concentrations of PCBs in Great Lakes fish.[14]	Baseline: (2000) 0%	34%	37%	40%	43%	46%

Principal Actions to Achieve Progress

New actions for FY 2010 to 2014 to protect the Great Lakes from toxic substances, clean up sediments and restore Areas of Concern include:

- Restore Areas of Concern/Remediate Contaminated Sediments — Accelerate the rate of sediment cleanup in AOCs throughout the Great Lakes basin through programs such as the Great Lakes Legacy Act, Water Resources Development Act, Natural Resource Damage Assessment, and Superfund. Restore and delist AOCs through strategic actions identified in Remedial Action Plans to restore individual beneficial uses. EPA will develop an enhanced management structure and ultimately increase our effectiveness in managing sediment remediation in AOCs. As other sediment sites may be identified, we will explore opportunities and authorities to address remediation at those sites.

Action Illustration: Contaminated Sediment Clean-up Can Also Provide Benefits to Environmental Justice Areas

Contaminated sediment is a significant source of fish contamination, which can disproportionately impact women of childbearing years, subsistence anglers and communities in areas of Environmental Justice concern. In 2009, the U.S. EPA Great Lakes National Program Office, Wisconsin Department of Natural Resources and other partners conducted a Great Lakes Legacy Act clean-up of heavily contaminated sediments in a section of Milwaukee's Kinnickinnic River. The project removed around 167,000 cubic yards of sediment contaminated with PCBs and polycyclic aromatic hydrocarbons, which were available to resident fish and had worked their way downstream to Lake Michigan, causing environmental harm and impeding commercial and recreational navigation. This project was in a section of the City of Milwaukee considered to be an area of potential Environmental Justice concern. Although the project was not carried out as an Environmental Justice project, nor was it in response to specific Environmental Justice concerns, the clean-up resulted in improved environmental health for the area, and it received support from local environmental groups and community health advocates. The clean-up also spurred economic revitalization efforts in this urban river community. Yet another benefit was that the clean-up provided job training opportunities to disadvantaged youth.

- Strategic Pollution Prevention and Reduction Projects — Prevent toxic and potentially toxic pollutants from entering the Great Lakes through a variety of new strategic actions, working closely with state, tribal and local governments. Initiate toxic reduction activities targeting mercury in emissions, products and waste, and expand Clean Sweep, and other collection programs and prevention practices to promote the safe disposal and elimination of pesticides, pharmaceuticals and other waste stream pollutants that can cause impairments.
- Protect Human Health through Safer Fish Consumption — Increase protection of Great Lakes fish consumers from harmful chemicals such as mercury and PCBs, with sound and sensible advice provided through enhanced and expanded state and tribal fish advisory programs. Work closely with the Great Lakes medical and health communities to educate the general public regarding the benefits and risks of Great Lakes fish consumption.
- Measuring Progress and Assessing New Toxic Threats — Measure progress in cleaning up toxics in the Great Lakes environment through comprehensive monitoring and assessment. Coordinate with efforts to update the Toxic Substances Control Act to ensure national programs take into account threats developing in the Great Lakes. Identify significant sources and impacts of new toxics to the Great Lakes ecosystem through robust surveillance as well as laboratory and field studies, in order to devise and implement effective control strategies.

Action Illustration:
Beneficial Use Impairments

AOC cleanup actions are directed toward restoring beneficial uses that have been impaired. A Beneficial Use Impairment (BUI) is defined in the Great Lakes Water Quality Agreement with Canada as "a change in the chemical, physical or biological integrity of the Great Lakes System sufficient to cause any of the following:

1) restrictions on fish and wildlife consumption;
2) tainting of fish and wildlife flavor;
3) degradation of fish wildlife populations;
4) fish tumors or other deformities;
5) bird or animal deformities or reproduction problems;

6) degradation of benthos;
7) restrictions on dredging activities;
8) eutrophication or undesirable algae;
9) restrictions on drinking water consumption, or taste and odor problems;
10) beach closings;
11) degradation of aesthetics;
12) added costs to agriculture or industry;
13) degradation of phytoplankton and zooplankton populations;
14) and loss of fish and wildlife habitat."

EPA cooperates with states to ensure actions are taken to address the BUIs in each AOC. When monitoring data shows a beneficial use is no longer impaired, the state and EPA can delist that BUI. When data shows all identified beneficial use impairments have been eliminated, then EPA can submit a proposal to the International Joint Commission (IJC) for removing the designation of AOC in its entirety from the area. If the IJC concurs that all BUIs have been restored, the parties, public and IJC can agree to delist the AOC as a whole.

Focus Area 2: Invasive Species

Problem Statement

Progress toward restoring the Great Lakes has been significantly undermined by the effects of non-native aquatic, wetland, and terrestrial invasive species. More than 180 aquatic nuisance species (ANS) now exist in the Great Lakes. The most invasive of these, including the well known zebra mussel, reproduce and spread, ultimately degrading habitat, out-competing native species, and short-circuiting food webs.

Prevention is the most cost-effective approach to dealing with organisms that have not yet arrived and could potentially threaten the lakes. New invasive species can be introduced into the Great Lakes region through various pathways, including commercial shipping, canals and waterways, trade of live organisms, and activities of recreational and resource users. Once invasive species establish a foothold in the Great Lakes, they are virtually impossible to eradicate; however, invasive species still need to be controlled to maintain the health of the Great Lakes ecosystem. Given that commercial shipping is a proven vector for invasive species transfer, advancing the development and use of ballast water treatment systems that are protective of fresh water is a high priority. Promising technology and innovative management practices that can significantly reduce the cost of control are under development. Control efforts will be accelerated in order to prevent the further spread of the organisms to inland lakes, the Mississippi River watershed, and beyond.

Long Term Goals

- Goal 1: The introduction of new invasive species to the Great Lakes basin ecosystem is eliminated, reflecting a "zero tolerance policy" toward invasives.
- Goal 2: The risk of introduction of species, which are imported for various uses, into the Great Lakes is minimized.
- Goal 3: The spread of invasive species, by means of recreational activities, connecting waterways, and other vectors, beyond their current range is prevented.
- Goal 4: A comprehensive pro gram for detection and tracking newly identified invasive species in the Great Lakes is developed and provides up-to-date critical information needed by decision makers for evaluating potential rapid response actions.
- Goal 5: An effective, efficient and environmentally sound pro gram of inte grated pest management for invasive species is developed and implemented, including pro gram functions of containment, eradication, control and miti gation.

Action Illustration:
Managing ANS already in the Great Lakes

While efforts to date have been unable to eradicate the sea lamprey from the Great Lakes, an ongoing program coordinated by the Great Lakes Fishery Commission has been able to control their populations. In 2007, the Sea Lamprey Control Program focused on nearly 50 streams in the basin, using a lampricide to eliminate over 5.5 million larval sea lampreys. Each parasitic phase sea lamprey has the capability of killing upwards of 40 pounds of lake trout during its year in the lakes. The successful control program continues to ensure sport fish rehabilitation and protects a fishery valued at over $7 billion in annual direct and indirect benefits.

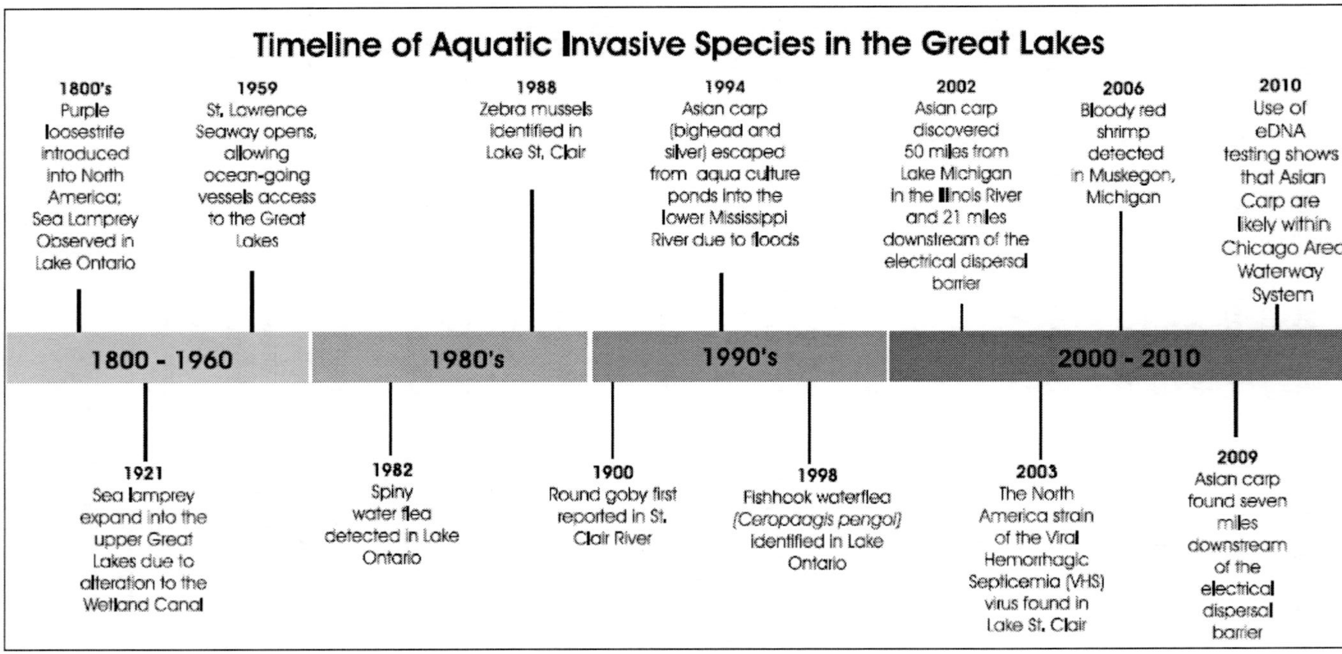
Timeline of Aquatic Invasive Species in the Great Lakes
1800's Purple loosestrife introduced into North America; Sea Lamprey Observed in Lake Ontario
1959 St. Lawrence Seaway opens, allowing ocean-going vessels access to the Great Lakes
1988 Zebra mussels identified in Lake St. Clair
1994 Asian carp (bighead and silver) escaped from aqua culture ponds into the lower Mississippi River due to floods
2002 Asian carp discovered 50 miles from Lake Michigan in the Illinois River and 21 miles downstream of the electrical dispersal barrier
2006 Bloody red shrimp detected in Muskegon, Michigan
2010 Use of eDNA testing shows that Asian Carp are likely within Chicago Area Waterway System
1800 - 1960
1980's
1990's
2000 - 2010
1921 Sea lamprey expand into the upper Great Lakes due to alteration to the Wetland Canal
1982 Spiny water flea detected in Lake Ontario
1900 Round goby first reported in St. Clair River
1998 Fishhook waterflea (Ceropaagis pengoi) identified in Lake Ontario
2003 The North America strain of the Viral Hemorrhagic Septicemia (VHS) virus found in Lake St. Clair
2009 Asian carp found seven miles downstream of the electrical dispersal barrier

Objectives

- By 2011, eight state ANS management plans will be established or revised to include rapid response capabilities. By 2014, eight state-based, multi-agency rapid response plans will be implemented and 22 mock exercises to practice responses carried out under those plans and/or actual response actions will be completed.
- Six technologies that prevent the introduction of invasive species and four technologies that either contain or control invasive species will be developed or refined and piloted by 2011. Ten technologies that prevent the introduction of invasive species and five technologies that either contain or control invasive species will be developed or refined and piloted by 2014.
- By 2011, methodology and protocols will be piloted for the coordinated monitoring methodology and shared protocols for basinwide invasive species surveillance. By 2014, a basinwide surveillance program with shared sampling protocols and methodologies to provide early detection of non-native species will be operational.
- By 2014, a 40 percent reduction in the yearly average rate of invasive species newly detected in the Great Lakes ecosystem will be achieved, compared to the period 2000-2009.
- By 2014, invasive species populations within the Great Lakes Ecosystem will have been controlled and reduced, as measured in populations controlled to a target level in 6,500 acres of managed area and by removing 5,000 pounds of invasive species from the Great Lakes ecosystem.

Action Illustration:
Early Detection and Rapid Response

In summer 2008, a mock rapid response exercise for aquatic invasive species was conducted in the Pennsylvania waters of Lake Erie. Local, state and federal partners responded to a simulated report of a Snakehead fish by a member of a local fishing club. This mock exercise allowed agencies to test the effectiveness and further refine their multi-jurisdictional rapid response plan. This work will be greatly expanded to all the Great Lakes to enhance coordination between management agencies. basinwide surveillance program will also be established by natural resource management and environmental protection agencies.

- By 2014, approximately 10 million recreation and resource users will be educated on best practices that prevent the introduction and spread of invasive species.

Measures of Progress

The Initiative will si gnificantly advance efforts to prevent new introductions of invasive species in the Great Lakes basin and to stop the further spread of invasive species in the Great Lakes basin.[15] Great Lakes Interagency Task Force agencies will work to further develop the initial set of measures by which pro gress will be evaluated in this focus area. While the Great Lakes community has pioneered many approaches to address invasive species, governmental pro grams and local efforts are still evolving to meet this threat. The GLRI will build upon and expand these pro grams, adapting our management approaches as we learn from our efforts. The measures by which pro gress will be evaluated in this focus area are:

Measure	Baseline/ Universe	2010 Target	2011 Target	2012 Target	2013 Target	2014 Target
1. Rate of nonnative species newly detected in the Great Lakes ecosystem.	Baseline: 1.3 species per year Universe: 181 species	1.3 species per year	1.1 species per year	1.0 species per year	1.0 species per year	0.9 species per year
2. Acres managed for populations of invasive species controlled to a target level. (cumulative)	Baseline: 0 acres	1,000 acres	1,500 acres	3,000 acres	4.500 acres	6,500 acres
3. Number multi- agency plans established, mock exercises to practice rapid responses carried out under those plans,	Baseline: 0 rapid response exercises/ actions	4 rapid response exercises/ actions; and 8 plans established	7 rapid response exercises/ actions	12 rapid response exercises/ actions	17 rapid response exercises/ actions	22 rapid response exercises/ actions

Measure	Baseline/ Universe	2010 Target	2011 Target	2012 Target	2013 Target	2014 Target
and/or actual rapid response actions (cumulative).						
4. Number of recreation and resource users contacted on best practices that prevent the introduction and spread of invasive species. (cumulative)	Baseline: 0 users	1,000,000 users	1,750,000 users	4,750,000 users	7,250,000 users	9,750,000 users

Principal Actions to Achieve Progress

Principal actions for FY 2010 to 2014 to prevent new introductions of ANS in the Great Lakes basin and stop the further spread of ANS include:

- Develop Ballast Water Treatment that Protects Freshwater Ecosystems — Develop a coordinated approach to the development of ballast water treatment suitable for freshwater ecosystems, though the use of laboratory, land-based and/or ship-board testing, and verification of treatment technologies in coordination with the maritime industry. Support work to reduce ship-mediated introductions through hull and anchor chain fouling.
- Implement Early Actions to Address Water Pathways Vectors — Identify key waterways that could introduce ANS to the Great Lakes and implement actions such as ecological separation to reduce this risk. Existing canals and extreme storm events can form hydrological connections that may introduce invasive species into the Great Lakes. Models and analysis of hydrological connections under different weather conditions are needed to identify and minimize risks of these connections.
- Prevention by Broad Stakeholder Outreach and Education — Promote actions, including coordinated education and outreach, which will prevent the introduction and spread of invasive species through recreational uses such as hunting, fishing and recreational boating.

- Develop and Demonstrate Innovative Control Technology — Promote the development and use of new control technologies, including biological control methods, which will significantly reduce the cost and/or increase the effectiveness of invasive species control measures.

Action Illustration:
Asian Carp: Using the G LRI to respond to ecological threats

The innovative use of environmental DNA (eDN A) sampling has revealed that Asian carp are likely within the Chicago Area Waterway System but at levels lower than may be detected using traditional sampling methods. This recent discovery mobilized governmental agencies to undertake a rapid response action in December 2009 to prevent the further migration of Asian Carp while repairs were made to an Electrical Barrier in the waterway. Subsequent eDN A analyses revealed that carp may be near the O'Brian Lock, Wilmette Lock, and Calumet Harbor, within the Great Lakes basin.

Great Lakes environmental protection and natural resource agencies have formed an Asian Carp Regional Coordinating Committee to respond to information as it becomes available. Their coordinated approach is described in the Asian Carp Control Strategy Framework. GLRI funding is helping to implement actions described in that Framework, including:

- U.S. Army Corps of Engineers installation of structures at the electric barrier site to reduce the risk of bypass during high water events,
- U.S. Fish and Wildlife Service and Illinois Department of Natural Resource efforts to detect and remove Asian Carp from the system, and
- U.S. Geological Survey efforts to develop innovative control technologies.

Agencies will continue to adapt their management strategies as new information becomes available. GLRI funding will continue to support these adaptive management efforts.

- Support States' Role in Invasive Species Prevention and Control — Support the development and on-theground implementation of ANS Management Plans for each of the Great Lakes states.
- Control Key Invasive Species and Investigate Causal Mechanisms by which ANS impact Native Species Develop a better understanding and models of ecosystem interactions and management options for minimizing the impact of ANS, including new treatment or control methods.
- Establish Early Detection and Rapid Response Capability — Work with government agencies to initiate surveillance activities to detect new ANS and establish the capacity, methods and contingency plans for a rapid response. Joint planning will allow the mobilization of shared resources to create the best opportunity for eradicating species before they become established.

No effort to pursue these goals in the Great Lakes can ignore the rapidly changing situation involving Asian Carp. The migration of Asian Carp through the Chicago Area Waterway System is the most recent and most acute ANS threat facing the Great Lakes today. This species has wreaked permanent havoc on the Mississippi River Basin, where the fish's rapid expansion of population and range has overwhelmed the river ecosystem by consuming plankton, a vital part of the food chain. As large populations of Asian carp have become established, the cumulative effects of these species have reduced food for native fishes, caused risks to human safety, and created impacts to regional economies that rely on fishing and boating.

The fish is now threatening similar damage to the ecosystem of the Great Lakes, as well as significant economic damage to the Great Lakes sport fishing industry. The Great Lakes community faces an urgency to prevent this threat from materializing.

An inter-governmental Regional Coordinating Committee has been established to oversee the implementation of an Asian Carp Control Strategy Framework, which is a cooperative effort by local, state, provincial, federal and even binational entities. The Framework is intended to provide direction for implementation efforts to prevent the establishment of ANS from migrating through artificially-connecting waterways that exist in the Chicago area, as well as other parts of the Great Lakes watershed, and if already in the basin, to work to implement efforts that might prevent further migration into the system.

The federal partners are working quickly to incorporate Carp and other ANS control efforts into the Initiative. This work pursues the objectives and measures of progress described in this section.

Focus Area 3: Nearshore Health and Nonpoint Source Pollution

Problem Statement

Nearshore and open waters provide drinking water for municipalities and habitat for numerous species of birds, fish and other aquatic life. This is the area in which most residents and visitors experience the Great Lakes through swimming, boating and other forms of recreation. Nearshore water quality has become degraded, as evidenced by eutrophication.[16] The environmental stressors causing these problems include excessive nutrient loadings from both point and nonpoint sources; bacteria and other pathogens responsible for outbreaks of botulism and beach closures; development and shoreline hardening that disrupt habitat and alter nutrient and contaminant runoff; and agricultural practices, which increase nutrient and sediment loadings. Additional shoreline stresses can be traced to failing septic systems, grey water pipes (pipes containing non-hazardous household substances like soap), inadequate pump-out stations for recreational boats, and even invasive species.

Nonpoint sources are now the primary contributors of many pollutants to the lakes and their tributaries. Although some nonpoint sources act on a whole-basin scale (e. g., atmospheric deposition of toxic substances), many smaller scale sources contribute to degraded water quality in Great Lakes tributaries and nearshore waters. Sediment is a si gnificant nonpoint pollution problem facing our lakes, rivers and streams. When soil gets washed into our waterways, it can smother fish habitats and also can carry pollutants that threaten water quality. The complexity of pollutants and their presence in soil, water and air make pollution abatement for nonpoint sources particularly difficult to address. Control strate gies to date have been inadequate to deliver the degree of stream and lake restoration necessary for the protection and maintenance of the Great Lakes. However, implementation of agricultural, urban stormwater and other watershed best management practices can have multiple benefits, including simultaneous reductions in runoff of soils, nutrients and pesticides, as well as protection of drinking water sources.

What is the "Nearshore"?

The aquatic nearshore can be considered to begin at the shoreline and extend offshore to the depth at which the warm surface waters typically reach the bottom in early fall, generally 20m - 30m deep, and terrestrial nearshore areas range from narrow beaches to inland features influenced by Great Lakes processes.

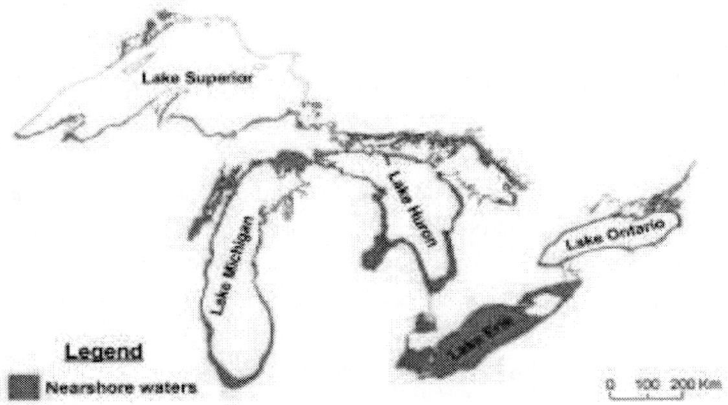

Long Term Goals

- Goal 1: Nearshore aquatic communities consist of healthy, self-sustaining plant and animal populations dominated by native and naturalized species.
- Goal 2: Land use, recreation and economic activities are managed to ensure that nearshore aquatic, wetland and upland habitats will sustain the health and function of natural communities.
- Goal 3: The presence of bacteria, viruses, pathogens, nuisance growths of plants or animals, objectionable taste or odors, or other risks to human health are reduced to levels in which water quality standards are met and beneficial uses attained to protect human use and enjoyment of the nearshore areas.
- Goal 4: High quality bathing beach opportunities are maintained by eliminating impairments from bacterial, algal and chemical contamination; effective monitoring for pathogens; effective modeling of environmental conditions, where appropriate; and timely communications to the public about beach health and daily swimming conditions.

- Goal 5: A significant reduction in soil erosion and the loading of sediments, nutrients and pollutants into tributaries is achieved through greater implementation of practices that conserve soil and slow overland flow in agriculture, forestry and urban areas.
- Goal 6: High quality, timely and relevant information about the nearshore areas is readily available to assess progress and to inform enlightened decision making.

Objectives

- By 2010, EPA will compile and map the highest priority watersheds for implementation of targeted nonpoint source pollution control measures.
- By 2014, remediation, restoration and conservation actions in at least one targeted watershed in each Great Lake basin will control erosion, reduce nutrient runoff from urban and agricultural sources, and improve habitat to protect nearshore aquatic resources.

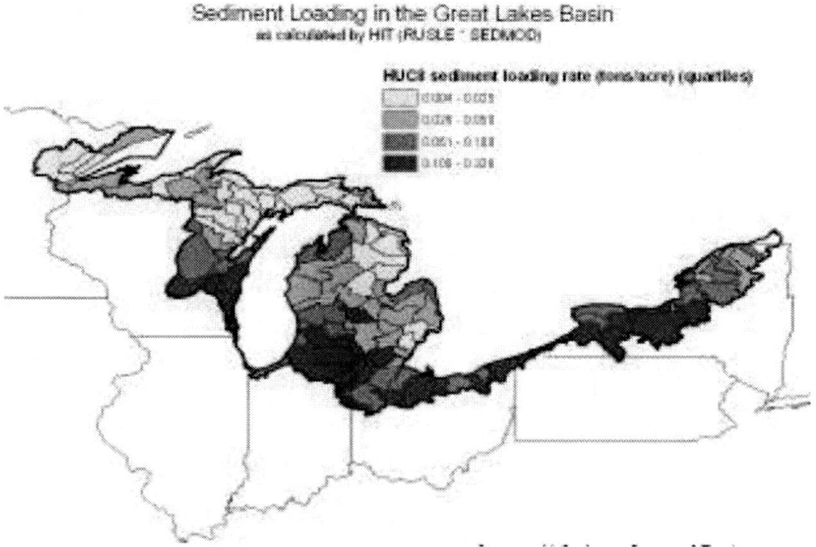

Source: U.S. Army Corps of Engineer.

- By 2014, a baseline will be established for total suspended solids loadings from targeted tributaries.
- By 2014, a measurable decrease will be achieved in soluble phosphorus loading from 2008 levels in targeted tributaries.

- By 2014, the causes of nutrient-related nearshore biolo gical impairments will be better understood, and following local or watershed remedial actions, the number and severity of incidences of harmful al gal blooms (HABs), avian botulism, and/or excessive Cladophora growth will be si gnificantly reduced from 2008 levels.
- By 2014, a comprehensive nearshore monitoring pro gram will have been established and implemented, including a publicly accessible reporting system, based on a suite of environmental indicators.
- By 2014, 50 percent of hi gh priority[17] Great Lakes beaches will have been assessed using a standardized sanitary survey tool to identify sources of contamination.

Action Illustration:
Targeted Geographic Initiatives

Because some Great Lakes subwatersheds show severe signs of stress, targeted geographical areas will receive focused efforts toward restoration activities, e.g., agricultural best management practices. Watersheds for the Genesee River, Grand Calumet River and Harbor, Green Bay/Fox River, Maumee River, St. Louis River, and Saginaw River, for example, are places where long term environmental problems have been clearly identified, and they are expected to be among those targeted for Initiative efforts. Watersheds associated with other AOCs may also be targeted to demonstrate more immediate results of actions to restore beneficial use impairments.

- By 2014, 20 percent of hi gh priority Great Lakes beaches will have begun to implement measures to control, manage or remediate pollution sources identified through the use of sanitary surveys.
- By 2014, rapid testing or predictive modeling methods (to improve the accuracy of decisions on beach postings to better protect public health) will be employed at 33 percent of high priority beaches.
- By 2014, the area of agricultural lands in conservation and/or utilizing conservation tillage practices will increase by 50 percent over 2008 levels.

Measures of Progress

The Initiative will si gnificantly improve the health of Great Lakes nearshore areas and will advance the reduction of nonpoint source pollution to levels that do not impair nearshore waters. The measures by which pro gress will be evaluated in this focus area are:

Measure	Baseline/ Universe	2010 Target	2011 Target	2012 Target	2013 Target	2014 Target
1. Five year average annual loadings of soluble reactive phosphorus[18] from tributaries draining targeted watersheds.	Baseline: 2003-7 Fox River: 212 Saginaw R: 133 Maumee R: 623 St. Louis R: **[19] Genesee R:85[20]	0% reduction Fox: 212 Saginaw: 133 Maumee: 623 St. Louis: ** Genesee: 85	0.5% Fox: 211 Saginaw: 131 Maumee: 617 St. Louis: ** Genesee: 85	1.5% Fox: 209 Saginaw: 130 Maumee: 611 St. Louis: ** Genesee: 84	2.5% Fox: 207 Saginaw: 129 Maumee: 604 St. Louis: ** Genesee: 83	4.5% Fox: 202 Saginaw: 126 Maumee: 592 St. Louis: ** Genesee: 81
2. Percentage of beaches meeting bacteria standards 95% or more of beach days.	Baseline: 86% (2006) Universe: 100%	86%	87%	87%	88%	89%
3. Extent (sq. miles) of Great Lakes Harmful Al gal Blooms. *21	Baseline: 2008 estimate TBD	0% reduction	4% reduction	7% reduction	8% reduction	12% reduction
4. Annual number of days U.S. Great Lakes beaches are closed or posted due to nuisance algae .22	Baseline: 200 beach days (estimate)	0% improvement: 200 days	4% improvement: 192 days	7% improvement: 186 days	8% improvement: 184 days	12% improvement: 176 days

Measure	Baseline/ Universe	2010 Target	2011 Target	2012 Target	2013 Target	2014 Target
5. Annual volume of sediment deposition in defined harbor areas in targeted watersheds (cu yards).[23]	Baseline: 2008 Toledo Harbor: 1 million cubic yards	0% improvement: 1 million cubic yards	1% improvement: 0.99 million cubic yards	1% improvement: 0.99 million cubic yards	2% improvement: 0.98 million cubic yards	2.5% improvement: 0.975 million cubic yards
6. Acres in Great Lakes watershed with USDA conservation practices implemented to reduce erosion, nutrients and/or pesticide loading under Farm Bill Pro grams.[24]	Baseline: 165,000 acres	2% increase 168,300 acres	10% increase 181,500 acres	25% increase 206,250 acres	40% increase 231,000 acres	50% increase 247,500 acres

Principal Actions to Achieve Progress

The principal actions to improve the health of nearshore areas and reduce nonpoint source pollution to levels that do not impair nearshore Great Lakes waters include:

- Targeting Watershed Plan Implementation - Significant progress has been made nationally and in the Great Lakes basin in addressing soil erosion and in reducing nutrient or other contaminant loads to tributaries to the Great Lakes through the existing state and federal programs. Water quality problems still exist, however, with loadings of sediment and nutrients still unacceptably high in a number of areas and degraded watershed conditions linked to impaired nearshore biological communities. This results in increased costs for navigation dredging of harbors, and in localized environmental problems such as mats of rotting algae on swimming beaches and along the shore. Initiative efforts in this area will address high priority watersheds, performing scientific analyses to strategically target where on-the-ground actions can be most effective, and providing supplemental funding to implement those actions. Innovative actions may include green infrastructure, low-impact development solutions, and containment of wastes from confined animal feedlot operations. This will involve close collaboration between state programs, Great Lakes tribal governments, the U.S. Department of Agriculture's Natural Resources Conservation Service and U.S. Forest Service; the U.S. Army Corps of Engineers; Department of Interior's U.S. Fish and Wildlife Service and U.S. Geological Survey; and EPA.
- Identify sources and reduce loadings of nutrients and soil erosion - These activities will contribute to the reduction or elimination of the number and severity of incidences of ecosystem disruptions, including Cladophora, HABs, botulism and other issues associated with eutrophication. Activities will include: applying research and modeling to identify the most effective actions to prevent incidences of Cladophora, HABs and botulism; enhancing or implementing watershed practices to reduce export of nutrients and soils to the nearshore waters; and establishing and implementing total maximum daily loads (TMDLs) for phosphorus, scaled from river reaches to watersheds.
- Improve Public Health Protection at Beaches - Humans are put at risk when exposed to pathogenic bacteria. These activities will reduce risk

to human health at swimming beaches by reducing the abundance of pathogenic organisms to levels below established criteria, increasing the effectiveness of monitoring for pathogens, modeling environmental conditions likely to result in elevated levels of bacteria, or enhancing communications to the public about daily swimming conditions.

- Generate Critical Information for Protecting Nearshore Health - The nearshore environment of the Great Lakes is highly varied, including relatively unspoiled shorelines, highly urbanized reaches, tributary mouths, bays, wetlands and other environmental features. These activities will promote the collection of data about nearshore conditions and stresses, the assessment of information and management implications, or the dissemination of information to all potential users in the Great Lakes community.

Focus Area 4: Habitat and Wildlife Protection and Restoration

Problem Statement

The health of Great Lakes habitats and wildlife depends upon the protection and restoration of ecosystems: the Great Lakes, the coastline, wetlands, rivers, connecting channels and watersheds. Humans benefit from healthy ecosystems. Healthy Great Lakes, for example, provide us with clean drinking water; rare wildlife populate a variety of unique coastal habitats; wetlands help control floodwaters; rivers transport sediments, nutrients and organic materials throughout the watershed; forests provide oxygen while reducing erosion and sedimentation; and, upland habitats produce topsoil and habitats for pollinators and bio-control agents. Fully resilient ecosystems buffer the impacts of potential problems such as climate change.

A multitude of threats affect the health of Great Lakes habitats and wildlife. Habitat destruction and degradation due to development; competition from invasive species; the alteration of natural lake level fluctuations due to artificial lake level management and flow regimes from dams, drain tiles, ditches, and other control structures; toxic compounds from urban development, poor land management practices and non-point sources; and, habitat fragmentation have impacted habitat and wildlife. This has led to an altered food web, a loss of biodiversity, and poorly functioning ecosystems. Many opportunities exist to protect and restore critical elements of the Great

Lakes ecosystem even as we strive to improve our understanding of emerging issues and their impacts.

Long Term Goals

- Goal 1: Protection and restoration of Great Lakes aquatic and terrestrial habitats, including physical, chemical, and biological processes and ecosystem functions, maintain or improve the conditions of native fish and wildlife.
- Goal 2: Critical management activities (such as stocking native fish and other aquatic species, restoring access of migratory fish species at fish passage barriers, and identifying and addressing diseases) protect and conserve important fish and wildlife populations.
- Goal 3: Sound decision making is facilitated by accessible, site specific and landscape-scale baseline status and trend information about fish and wildlife resources and their habitats.
- Goal 4: High priority actions identified in strategic plans (such as state and federal species management, restoration and recovery plans, Lakewide Management Plans, Remedial Action Plans, and others) are implemented, lead to the achievement of plan goals, and reduce the loss of fish and wildlife and their habitats.
- Goal 5: Development activities are planned and implemented in ways that are sensitive to environmental considerations and compatible with fish and wildlife and their habitats.

Objectives

- By 2014, 4,500 miles of Great Lakes rivers and tributaries will be reopened and 450 barriers to fish passage will be removed or bypassed.
- By 2014, 82% of recovery actions for federally listed priority species will be implemented.
- By 2014, 53 percent of populations of native aquatic non-threatened and endangered species are self sustaining.
- By 2014, 97,500 acres of wetlands, wetland-associated uplands, and high priority coastal, upland, urban, and island habitats will be protected, restored or enhanced.
- By 2014, 100 percent of U.S. coastal wetlands in the Great Lakes basin will be assessed.
- By 2014, 30 habitat-related beneficial use impairments will be delisted across the Areas of Concern.

Action Illustration:
Lake Sturgeon Recovery

A resident of the Great Lakes for 10,000 years, the prehistoric lake sturgeon (Acipenser fulvescens) is more than a holdover from the distant past—it is a barometer of the health and diversity of the entire Great Lakes ecosystem. The largest fish in the Great Lakes, sturgeon can grow to be eight feet long and weigh more than 200 pounds. Remaining populations represent only a fraction of their former numbers and are now protected in most waters of the Great Lakes. Only 5 rivers basin-wide have an annual spawning run of 200 or more adults and only one river exceeds the minimal viable population spawning run target of 350 adults. Major contributors to the sturgeon's decline are historical over-harvesting and habitat loss resulting from the damming of tributary waters, siltation resulting from deforestation, poor agricultural practices and dredging. Pollution from nutrients and contaminants have also reduced water quality and hindered reproductive success. The sturgeon's late maturity and infrequent spawning has also contributed to its limited recovery.

Over 40 partnerships have been operating throughout the Great Lakes to conserve, protect, and enhance lake sturgeon populations through numerous cooperative efforts. Great Lakes Restoration Initiative funding will now accelerate sturgeon rehabilitation in the basin through on-the-ground habitat enhancement and restoration projects, such as restoring fish passage to historical spawning areas, as well as supporting the rearing, stocking, and assessment of sturgeon populations. Initiative funding will also support a coordinated, multi-agency investigation of environmental contaminants in sturgeon and will improve evaluation and prioritization of habitat restoration projects.

Of the 26 tributaries that support sturgeon in the Great Lakes, 16 of these will benefit from either initiation of stream-side rearing facilities to enhance recruitment or habitat enhancement through the GLRI. As a result of enhanced sturgeon management and rehabilitation efforts through the GLRI, a basin-wide total of 25,000 fingerling lake sturgeon will be stocked to enhance existing small but remnant populations. A boosted stocking effort through FY2014 is expected to promote lake sturgeon spawning success and will promote populations toward self-sustaining levels. Habitat assessment and enhancement projects will be implemented on 20 existing sturgeon streams, and fish passage will be provided on 2 barriers that limit sturgeon access to upstream habitat.

Improvements to the health of lake sturgeon will be captured in the fifth measure of progress below: "Percent of populations of native aquatic non-threatened and endangered species self sustaining in the wild."

Measures of Progress

The Initiative will significantly accelerate Great Lakes habitat and wildlife protection. The measures by which progress will be evaluated in this focus area are:

Measure[25]	Baseline/ Universe	2010 Target	2011 Target	2012 Target	2013 Target	2014 Target
Miles of rivers reopened for fish passage.	Baseline: 0 Universe: 20,000 miles	1,000 miles	1,500 miles	2,500 miles	3,500 miles	4,500 miles
Number of fish passage barriers removed or bypassed.	Baseline: 0 Universe: 5,000 barriers	100 barriers	150 barriers	250 barriers	350 barriers	450 barriers
Number of species delisted due to recovery	Baseline (2009): 0 species Universe: 28 listed species- Great Lakes	0	1[26]	1	1	2
Percent of recovery actions implemented for priority listed species	Baseline (2009): 0 Universe: 414 recovery actions27	16% 68/414	33% 138/414	51% 211/414	67% 277/414	82% 338/414
% of populations of native aquatic non-threatened and endangered species self-sustaining in the wild.	Baseline (2009): 27% 39/147 populations[28] Universe: 147 populations	33% 48/147 populations	35% 52/147 populations	41% 60/147 populations	47% 69/147 populations	53% 78/147 populations
Number of acres of wetlands and wetland-associated uplands protected, restored and enhanced.	Baseline: 0 Universe: 550,000 acres	5,000 acres	7,500 acres	12,500 acres	17,500 acres	22,500 acres

Measure[25]	Baseline/ Universe	2010 Target	2011 Target	2012 Target	2013 Target	2014 Target
Number of acres of coastal, upland, and island habitats protected, restored and enhanced.	Baseline: 0 Universe: 1,000,000 acres	15,000 acres	20,000 acres	35,000 acres	50,000 acres	75,000 acres
% of U.S. coastal Great Lakes wetlands assessed.	Baseline: 0 Universe: 100%	20%	40%	60%	80%	100%
Number of habitat- related BUIs removed from the 27 U.S. AOCs so impaired[29].	Baseline: 4 Universe: 75 so impaired	9 BUIs removed	12 BUIs removed	18 BUIs removed	24 BUIs removed	30 BUIs removed

Principal Actions to Achieve Progress

The principal FY 2010-2014 actions to protect and restore habitat and wildlife include the following. Agencies will work together with states, tribes, municipalities, non-governmental organizations and industry to:

- Improve Aquatic Ecosystem Resiliency — Protect and restore aquatic habitats for fish and wildlife populations by reconnecting habitats through corridors to enhance biological diversity, reducing sediment and nutrient inputs, restoring natural hydrological processes, improving water quality, restoring ecosystem services, and increasing populations of native fish and wildlife through coordinated management actions.
- Maintain, Improve or Enhance the Populations of Native Species Implement restoration actions identified in species recovery and management plans; quantify habitat needs for depleted migratory bird species; propagate lake trout, coaster brook trout, lake sturgeon and other similar fingerlings for suppressed fish populations, assessing fish populations; and protect and restore culturally significant species.

Action Illustration:
Great Lakes Piping Plover Recovery

The Great Lakes population of piping plover (Charadrius melodus) is one of the region's most critically endangered species. As a result of long-term management and protection activities, reproductive success has improved, total population numbers have increased, and breeding pairs have increased from 17 in 1986 to 71 in 2009. However, the piping plover remains extremely vulnerable to extinction from threats including breeding site disturbances, habitat loss due to shoreline development and recreational use, adult mortality from predation and disease, and environmental events such as severe weather and oil spills. The U.S. Fish and Wildlife Service currently leads a multistate, binational recovery program that includes partners from other federal and state resource agencies, non-profit organizations, universities, zoos and volunteers. Funding through the GLRI will effectively boost partner participation with recovery efforts, as well as support the following goals over the next five years:

- Identify, protect and monitor each breeding area with a goal of increasing population levels approximately 15 percent, while sustaining wild reproductive success levels at 1.5 or more fledged chicks per adult pair.
- Implement population augmentation strategies including a salvage captive rearing program for abandoned eggs and chicks to increase overall reproductive success to greater than 1.7 fledged chicks per adult pair.
- Develop and implement an adaptive approach to predator management in an effort to sustain adult survival rates of 75 percent or greater.
- Conduct invasive species control efforts at 8-10 sites over a five year period to improve/preserve piping plover breeding habitat.
- Work with other agencies and non-profit organizations to secure conservation easements on three to four important piping plover breeding sites on private land.
- Support scientific research on plover ecology, population demographics and genetics.

> Improvements to the health of piping plover populations will be tracked through recovery actions implemented (see measure four).These efforts will significantly improve the chances for long-term recovery of the piping plover in the Great Lakes.

- Enhance Wetlands, Wetland-Associated Uplands, and High Priority Coastal, Upland and Island Habitats — Protect, restore, or enhance habitats by acquiring properties that are important to sustain fish and wildlife populations, restoring natural hydrological regimes, improving water quality, and restoring the chemical, physical, and biological integrity of ecosystems in each Great Lake basin.
- Identify, Inventory, and Track Progress on Great Lakes Habitats, Including Coastal Wetlands Restoration Assess progress toward restoring Great Lakes habitats by establishing baseline conditions and tracking trends; highlight the importance of coastal wetland conservation and restoration by implementing a longterm coastal wetland monitoring program and enhancing the National Wetlands Inventory.
- Restore Habitat Functioning in Areas of Concern — Improve habitats in degraded urban environments and Areas of Concern where beneficial use impairments limit ecosystem functioning by restoring habitats for native species populations and removing or isolating contaminants.

Focus Area 5: Accountability, Education, Monitoring, Evaluation, Communication and Partnerships

Problem Statement

The Action Plan requires additional oversight, monitoring, assessment, and coordination to succeed. Although the initiative benefits greatly from the years of scientific effort in the Great Lakes, there are still gaps in efforts to measure and monitor key indicators of overall ecosystem function, to evaluate restoration progress, and to provide the information decision makers need.

This information needs to be based on best available science, and compiled and communicated consistently to decision makers to allow them to assess ecosystem conditions and to track restoration progress. Efforts under the Initiative will establish the foundation for routine and consistent comprehensive ecosystem assessments. A comprehensive and efficient accountability system and well-defined metrics must be established to track progress under this Action Plan. This system needs to track progress of individual grant recipients and partner-led initiatives, as well as overall progress in meeting the short- and longterm goals and objectives of the Initiative as a whole.

This focus area is the necessary backbone for success in the others. There are gaps in knowledge and an inadequate understanding of complex and emerging issues such as nearshore dynamics and climate change impacts. This means new investment is needed, particularly in the first year of the GLRI, to set up the systems and tools necessary to understand, coordinate, teach, direct progress and capture results. Once these systems and tools are developed and in use, we anticipate that this focus area will decrease in significance as implementation increases in the other focus areas.

Outreach and education are also crucial in the effort to restore the Great Lakes. If the change we seek is to leave the Great Lakes better for the next generation than the condition in which we inherited them, then we must educate upcoming decision makers so tomorrow they are capable of learning from and extending today's restoration efforts. The growth of service learning programs is an encouraging sign that school systems are placing an increasing value on stewardship of and hands-on learning about the Great Lakes. A handful of Great Lakes curricula exist that satisfy state and other relevant learning standards, yet these rarely find their way into core education objectives for Great Lakes schools. We must continue to reach traditional classroom teachers and nature centers, but also reach non-traditional education outlets such as homeschoolers, national lakeshore and state park visitor centers, zoos, aquariums, and museums. Moreover, while there will always be some demand for the development of new curricula, alarmingly few resources are available today by the funding community for executing existing high-quality education specifically centered on the concept and practice of Great Lakes "restoration" that can show demonstrable results. Additionally, current education efforts often lack coordination, duplicate efforts and fail to move the Great Lakes education community toward collaborative best practices supportive of Great Lakes restoration. Support is especially lacking for the sustained implementation of existing high-quality standardized curricula through, for example, teacher training, funding for field trips, support for teachers' professional development opportunities and efforts focused on underserved communities, among other activities.

Action Illustration:
Comprehensive Great Lakes Coastal Assessment

The nearshore environment of the Great Lakes is highly varied, including relatively unspoiled shorelines, highly urbanized reaches, tributary mouths, embayments, wetlands and other environmental features. Regular, comprehensive monitoring has been difficult because there are over 10,000 miles of Great Lakes shoreline and effective technologies have only recently been developed. This complexity has inhibited the ability to track issues impacting human and ecological health and requires new approaches to monitoring that can take advantage of new technological tools. In keeping with national efforts to provide statistically valid assessments of water resources, a nearshore monitoring program in the Great Lakes will be implemented, which builds on ongoing long-term efforts and initiates new efforts focused on poorly characterized areas or elements. The nearshore monitoring program will provide the necessary scientific basis to assess the physical, chemical and biological integrity of the nearshore environment and to target restoration and protection efforts. The assessment will establish baseline conditions of environmental quality and variability of the nearshore waters, bottom substrate and biota.

The nearshore monitoring program will utilize a probability-based design for site selection, as well as for shipboard surveys targeted for specific purposes. It will also utilize remote sensing and other observing tools to help monitor the coastal areas. The program will include rivers that flow directly into the Great Lakes and drain watersheds greater than 250 square miles. River monitoring will allow calculation of seasonal and annual loadings from the watersheds to the Great Lakes.

EPA's National Coastal Condition Assessment will provide the basis for the initial effort by supplying a framework to which additional monitoring can be added. The additional monitoring will be an expansion of sites in the open waters, nearshore areas, and rivers using a probabilistic sampling design which allows statements on the overall quality of the Great Lakes coasts and also of the nearshore areas of individual lakes.

The Great Lakes span many different government jurisdictions along with their regulatory agencies and authorities: two countries, eight U.S. states, two Canadian provinces, 83 U.S. counties, thousands of cities and towns, 33 U.S.

tribal governments and more than 60 recognized First Nations in Canada. Through the 1909 Boundary Waters Treaty with Canada, the related Great Lakes Water Quality Agreement, and a host of other institutional arrangements, this region has a long history of governments at all levels working in partnership to protect and restore the Great Lakes. Federal coordination efforts have been greatly improved through efforts of the Great Lakes Interagency Task Force and its Regional Working Group. Binational efforts continue with help from the International Joint Commission, and through the Binational Executive Committee, which coordinates binational implementation of the provisions of the Great Lakes Water Quality Agreement.

These partnerships must continue and be further strengthened to address the complex issues faced by the Great Lakes.

Long Term Goals

- Goal 1: A cooperative monitoring and observing system provides a comprehensive assessment of the Great Lakes ecosystem.
- Goal 2: The necessary technology and programmatic infrastructure supports monitoring and reporting, including Great Lakes Restoration Initiative project deliverables by all agencies and participating stakeholders. Data and information are provided in reports that are public friendly, timely and available on the Internet. Reports present integrated and scaled data from watersheds to lakes to Great Lakes basinwide.
- Goal 3: Increase outreach and education for the Great Lakes, and provide ongoing K-12 education for students to understand the benefits and ecosystem functions of the Great Lakes so they are able to make decisions to ensure that restoration investments are enhanced over time.
- Goal 4: Expand the range of opportunities for Great Lakes stakeholders and citizens to provide input to the governments and participate in Great Lakes issues and concerns.
- Goal 5: Work under the goals and objectives of the Great Lakes Water Quality Agreement is coordinated between the U.S. and Canada through Lakewide Management Plans (LaMP) and other binational processes, programs, and plans.

Objectives

- By 2011, opportunities for collaboration, planning, data accessibility and accountability will be increased through the expanded use of internet-based technology.
- By 2011, an Accountability System will be developed and implemented for the Initiative. The system will integrate and make transparent strategic planning, budgeting and results monitoring.
- By 2011, a satellite remote sensing program will be implemented to assess Great Lakes productivity and biological (e. g., algal bloom) events.
- By 2011, outreach and education efforts are increased, including identifying and revising existing curricula to incorporate sustainable education needs for the Great Lakes that meet state and other relevant learning standards.
- By 2011, a refined suite of science-based indicators for development of a comprehensive assessment of Great Lakes ecosystem health will be identified, monitoring programs for those indicators will begin to be implemented, and restoration and protection actions tied to those assessments and programs assured.
- By 2011, social media access opportunities for basinwide public involvement in the Initiative will be in place.
- By 2012, education efforts under existing curricula that meet state and other relevant learning standards will be coordinated across states, and a system for tracking student and teacher outreach (quantitatively and qualitatively) for their use.
- By 2012, improved coordination with Canada will take place for pro grams under the Great Lakes Water Quality Agreement, particularly under the LaMPs, which will result in the achievement of 5-10 priority LaMP goals and actions.
- By 2014, a statistically valid and comprehensive assessment, using a probability-based desi gn, of Great Lakes water resources, will be established. The system will inte grate shipboard monitoring, remote sensing, automated sampling, and other monitoring or observing efforts. By 2016, the system will be in place for all of the Great Lakes and capable of providing a scientifically justifiable assessment of Great Lakes water resources.
- By 2014, timely data and information will be provided to decision makers at multiple scales within a framework of established baselines, targets, indicators of pro gress, and monitoring.

Action Illustration: What is a LaMP?

A Lakewide Management Plan, or "LaMP," is a plan of action to assess, restore, protect and monitor the ecosystem health of a Great Lake. It is used to coordinate the work of all the governmental and non-government partners working to improve a Lake's ecosystem. A public consultation process helps include attention to public concerns. Priority projects, such as sampling tributaries for chemicals, developing watershed plans, restoring habitats, conducting hazardous waste collections, and many others, are implemented in pursuit of LaMP goals and objectives. Many such projects are complex and time consuming, involving many partners and logistical concerns.

Measures of Progress

The Initiative will significantly improve collaborative Great Lakes decision making, transparency, and accountability for Great Lakes information. Representatives of the federal agencies below will work to gether to determine which existing agency inventory and monitoring data can be used to establish baselines for the various performance goals and to identify needed additional research and monitoring, outreach, and implementation. The measures by which pro gress will be evaluated in this focus area are:

Measure	Baseline/ Universe	2010 Target	2011 Target	2012 Target	2013 Target	2014 Target
Improvement in the overall aquatic ecosystem health of the Great Lakes using the Great Lakes 40-point scale.[30]	Baseline: 20 points Universe: 40 points	23 points	23.4 points	23.9 points	24.3 points	24.7 points
Number of priority LaMP projects that are completed.	Baseline: 0 projects	10 projects	12 projects	15 projects	18 projects	20 projects
Number of educational institutions incorporating new or existing Great Lakes protection and stewardship criteria into their broader environment education curricula.[31]	Baseline: 0	0 institutions	2 institutions	6 institutions	10 institutions	16 institutions

Principal Actions to Achieve Progress

The principal actions for FY 2010 to 2014 to achieve significant, measurable objectives include:

- Develop Great Lakes Restoration Accountability System — Develop and implement a transparent accountability system for the Great Lakes Restoration Initiative, including easy access to information and linkages to planning, budgeting, and results. With our new accountability system, partner agencies will report quality controlled information regularly on Initiative progress in meeting the objectives and targets of this Action Plan.
  Progress will also be reported with and through the LaMPs on the Great Lakes as a whole, as well as on each of the Lakes and Connecting Channels, and public forums will be harnessed to assist with the transfer and dissemination of information to the public.
- Measure and Evaluate the Health of the Great Lakes Ecosystem using the best available science — Enhance existing programs that measure and assess the physical, biological, and chemical integrity of the Great Lakes, including the Connecting Channels. Develop and implement a statistically valid assessment, using a probability-based design, of Great Lakes water resources, coinciding with intensive cooperative science and monitoring efforts for the Lakes. Implement strategic components relevant for Great Lakes decision making of the U.S. contribution to the Integrated Earth Observation System and the Integrated Ocean Observing System as part of the Global Earth Observing System of Systems. Develop a coordinated federal approach to address on the key scientific priorities needed to fully assess the impacts climate change may have on the health of the Great Lakes ecosystem and better manage those impacts. Promote the development and implementation of science-based indicators that will better assess and provide a better measure of accountability of actions to improve the health of the Great Lakes ecosystem.
- Support Great Lakes restoration education — Develop coordination mechanism(s) for Great Lakes restoration education efforts that meet state and other relevant learning standards. Identify the suite of participating educators and facilities (e. g., formal classroom teachers, homeschoolers, public and private community service learning programs, non-traditional education outlets such as nature centers, national lakeshore visitor centers, zoos, aquariums, museums, and

urban community centers, etc.). The number of educators who participate in teacher trainings and, as a result the number of participating students, is increasing. These trends should be made to translate into more institutions incorporating Great Lakes lessons into their curricula. Ensure metrics are included the Great Lakes Restoration Accountability System.

- Enhance Partnerships — Enhance coordination and collaboration among Great Lakes partners to help ensure that actions, projects and programs under the Great Lakes Restoration Initiative are efficient, effective and in furtherance of the U.S. - Canada Great Lakes Water Quality Agreement. Partnerships will be advanced and resources and capabilities leveraged through existing collaborative efforts such as the U.S. - Canada Binational Executive Committee, the State of the Lakes Ecosystem Conference, the U.S. - Canada Great Lakes Binational Toxics Strategy, Lakewide Management Plans, Four Agency Agreements, the Coordinated Science Monitoring Initiative, and Great Lakes Fisheries management. This principal action will also stress building ownership and a sense of joint responsibility among agencies, institutions and the public across the Great Lakes basin. Partners and stakeholders will be encouraged to suggest refinements to the Initiative, to monitor and judge progress, and to help adjust objectives and actions as progress is made. Existing mechanisms, such as the U.S. Policy Committee, will be used and new ones developed, if needed, to enable that participation.

CONCLUSION

If we care for the Great Lakes, they will continue to care for us — economically, ecologically and socially. Study after study, however, shows a Great Lakes in peril. These studies point the way to how we can work to make the ecosystem more resilient over time by reducing toxic pollution and cleaning up Areas of Concern, instituting a zero tolerance policy toward aquatic nuisance species, rehabilitating fish and wildlife habitat, shielding nearshore health from polluted "nonpoint source" runoff, and enhancing these efforts through proper accountability mechanisms, education and other measures. By declaring Great Lakes restoration a national priority with the proposal for significant new resources to address these issues, President Barack Obama is encouraging federal, state, municipal, tribal and civic representatives — all of us — to unify for leaving the Great Lakes better for the next generation.

ACRONYMS

ANS	Aquatic Nuisance Species
AOC	Area of Concern
BUI	Beneficial Use Impairment
EPA	U.S. Environmental Protection Agency
GLRC	Great Lakes Regional Collaboration
GLRI	Great Lakes Restoration Initiative
GPRA	Government Performance and Results Act
HAB	Harmful Algal Bloom
LaMP	Lakewide Management Plan
RFP	Request for Proposals
SRF	State Revolving Fund

End Notes

[1] The Initiative collapses these eight focus areas into five. A ninth focus area, concerning water quantity issues, was excluded from the GLRC Strategy because it was being addressed through a separate process: the Great Lakes-St. Lawrence River Basin Water Resources Compact. The Sustainable Development area also identified water infrastructure needs in the Basin. Infrastructure needs are not being addressed by GLRI but through increased funding for the State Revolving Funds through EPA's Regional offices.

[2] The FY 2010 plan, together with documents describing Agency actions therein and the programs and projects to carry it out, are posted to http://www.epa. gov/glnpo/glri/index.html.

[3] Objectives and targets in this plan are premised on an assumption that $300 million will be appropriated in FY 2011, and $475 million in subsequent years.

[4] Regional goals for water and sewer infrastructure improvements, which are supported by funding under the Clean or Drinking Water State Revolving Fund programs, are being addressed through those and other processes and are not part of this Action Plan.

[5] Note that these funds will not be directed toward water infrastructure programs that are addressed under the Clean Water or Drinkin Water State Revolving Fund program.

[6] The term "grants," as used in this document, includes both grants and cooperative agreements.

[7] As in previous years, some noncompetitive funding will be made available to states and tribes to ensure the capacity to participate in ongoing work to implement Lakewide Management Plans and Remedial Action Plans for Areas of Concern.

[8] Some long-term goals may not be achieved within the time span of this action plan.

[9] Baseline represents the starting point for the measure. Universe represents all that is likely possible to protect, restore, enhance, etc.

[10] This is an existing measure under the Government Performance Results Act.

[11] These projects represent on-the-ground actions that are being implemented in order to remove BUIs. For example, sediment removals, Superfund cleanups, habitat projects and others.

[12] Baseline represents the starting point for the measure. Universe represents all that is likely possible to protect, restore, enhance, etc.

[13] This is an existing measure under the Government Performance Results Act.

[14] The annual decline is 5% per year. This is based on an existing measure under the Government Performance Results Act. In FY 2010, 2008 data is compared to 2000; in FY 2011, 2009 data is compared to 2000; and so forth. PCBs are one indicator for a broader suite of persistent toxic substances and one of a number being tracked.

[15] While the regulatory tools of various agencies will also help advance these efforts, they are not currently a focus of this plan.

[16] The process by which a water body is enriched by nutrients such as phosphorus, resulting in excessive growth of algae, depletion of dissolved oxygen, and other impacts, including beach closings. In the nearshore zones of the Great Lakes, eutrophication has sometimes been manifested by harmful algal blooms; the green algae Cladophora washing ashore to make unsightly, odiferous rotting mats on beaches; and avian botulism.

[17] Beaches that the states identify as most frequently used and/or that have the highest risk. In 2008, there were 356 high priority beaches out of a total of 1,411 total beaches in the U.S. Great Lakes.

[18] Metric tons per year. Total phosphorus will also be measured. Targeted watersheds will receive focused efforts toward restoration activities, e. g., agricultural best management practices.

[19] **Calculations in progress, to be delivered 2010.

[20] Data for Genesee River are from 2006 and 2007 only.

[21] Biological responses to nutrients loadings are also dependent on other factors such as water temperature, timing and intensity of precipitation, and hydrologic features. Year-to-year variability in these features may mask local improvements in nutrients management. Satellite imagery may provide data for days during which HABs are reported by shoreline observers or boaters.

[22] This metric will be added to national surveys for beach managers for 2010. Nuisance algae can include Cladophora, HABs or other species, all of which are believed to be aggravated by elevated levels of phosphorus in the water.

[23] USACE dredges the federal shipping channel at Toledo Harbor each year. This area receives the highest rate of sedimentation in the Great Lakes, coming from the Maumee River watershed. Even small improvements in the rate of sedimentation here would reflect considerable efforts in the watershed to reduce erosion and farm runoff. Alternately, USACE conducts bathymetric surveys of commercial harbors each year, from which the volume of new fluvial sediment can be calculated for targeted watersheds. Because USACE does not dredge every location of every harbor each year, the estimated accumulation from a designated area over time will reflect the relative amount of sediments deposited from the tributary. This approach is currently in development.

[24] This measure reflects annual (not cumulative) implementation of conservation practices (from the Environmental Quality Incentives Program and Conservation Technical Assistance program) that will contribute to long term improvements of the listed outcomes.

[25] Out year targets for these measures are cumulative. The Universe represents all that is likely possible to protect, restore, enhance; baseline represents the number of acres etc. that are

already protected, restored, enhanced. All measures of progress included here are interim figures until final baselines are established.

[26] Lake Erie Watersnake 2011, Pitcher's thistle 2014.

[27] Numerator: # recovery actions implemented for Great Lakes priority listed species. Denominator: Total recovery actions for Great Lakes priority listed species, as defined in species recovery plans. Note that many recovery actions are implemented annually (i.e., update landowner records, monitor current populations, evaluate threats, etc.). Recovery actions are implemented until the threshold for action "completion" is met, as defined in each species recovery plan.

[28] Numerator: # of populations of native aquatic non-T&E and non-candidate species that are self-sustaining in the wild. Denominator: total # of native aquatic non-T&E and non-candidate populations.

[29] Also captured under Measure 2, Focus Area 1.

[30] This is an existing measure under the Government Performance and Results Act. The Great Lakes Index uses select Great Lakes ecosystem indicators (i.e., coastal wetlands, phosphorus concentrations, AOC sediment contamination, benthic health, fish tissue contamination, beach closures, drinking water quality, and air toxics deposition) and is based on a 1 to 5 rating system for each indicator, where 1 is poor and 5 is good. Improvements in the index and measures would indicate that fewer toxics are entering the food chain; ecosystem and human health is better protected; fish are safer to eat; water is safer to drink; and beaches are safer for swimming.

[31] Educational institutions include: state departments of education, primary and secondary school districts, colleges, universities, zoos, aquaria, museums, and nature/resource centers. Curricula will meet relevant official standards.

INDEX

A

abatement, 152
abuse, 123
access, 106, 108, 109, 112, 160, 161, 169, 171
accessibility, 108, 169
accountability, vii, 2, 5, 7, 21, 38, 39, 85, 101, 108, 111, 122, 130, 132, 136, 166, 169, 170, 171, 172
accounting, 59
acquisitions, 127
adaptation, 79, 80, 114
adults, 161
adverse event, 79
aesthetics, 143
age, 77, 79
agriculture, 32, 104, 143, 154
air toxics, 113, 117, 175
algae, 33, 105, 110, 117, 143, 156, 158, 174
ANS, 131, 144, 145, 147, 149, 151, 152, 173
AOC, 26, 125, 140, 141, 143, 173, 175
appropriations, viii, 2, 5, 7, 8, 9, 11, 15, 18, 26, 44, 45, 111, 135
Appropriations Act, 33
aquaria, 117, 175
aquatic habitats, 3, 163
aquatic life, 110
aromatic hydrocarbons, 142
Asia, 109
assessment, 21, 39, 48, 49, 51, 56, 58, 61, 73, 107, 108, 109, 122, 143, 159, 161, 165, 167, 168, 169, 171
assets, 28
atmospheric deposition, 76, 79, 114, 137, 152
attachment, 67
audit, 35, 91
authority(s), viii, 2, 5, 7, 13, 15, 46, 111, 134, 140, 142, 167
avian, 105, 155, 174
awareness, 138

B

bacteria, 66, 104, 105, 113, 152, 153, 156, 158
banks, 70
barriers, 6, 39, 50, 70, 106, 107, 127, 160, 161, 162
base, 8, 15, 17, 18, 35, 56, 88, 89, 90, 108, 109, 122, 135, 140, 167, 169, 171
beaches, ix, 48, 105, 110, 111, 113, 117, 121, 124, 126, 153, 155, 156, 158, 159, 174, 175
benefits, viii, 3, 18, 23, 29, 32, 37, 56, 71, 73, 85, 90, 108, 143, 145, 152, 165, 168
bias, 123, 134
biodiversity, 126, 130, 159

biological control, 39, 150
biological processes, 6, 106, 160
birds, 70, 152
blueprint, 62
botulism, 105, 152, 155, 158, 174
breeding, 164
Bureau of Indian Affairs, 10, 40, 110, 115
businesses, 77, 138

C

canals, 144, 149
cash, 111
challenges, 72, 81, 83, 85, 125, 127, 128, 131
chemical(s), vii, 1, 5, 6, 25, 36, 53, 87, 101, 104, 106, 125, 126, 128, 137, 138, 139, 143, 153, 160, 165, 167, 170, 171
Chicago, 79, 90, 114, 150, 151
children, 125
city(s), ix, 33, 78, 114, 121, 131, 142, 167
citizens, 26, 108, 168
clarity, 11, 55
classroom, 166, 171
classroom teacher(s), 166, 171
Clean Water Act, 131
cleaning, 4, 34, 143, 172
cleanup, 39, 121, 122, 125, 140, 142, 143
climate, vii, 1, 20, 31, 37, 59, 71, 74, 76, 78, 79, 80, 83, 84, 113, 114, 123, 124, 125, 126, 137, 159, 166, 171
climate change, 20, 31, 37, 59, 74, 76, 78, 79, 80, 83, 84, 113, 114, 123, 126, 137, 159, 166, 171
climatic factors, 52
closure, 90
coal, 53
Coast Guard, 10, 40, 66, 110, 115
coding, 90
collaboration, 4, 6, 11, 14, 15, 89, 108, 121, 128, 134, 135, 158, 169, 172
colleges, 117, 175
commercial, 36, 117, 142, 144, 174
communication, vii, 2, 5, 38, 39, 41, 101
community(s), 58, 64, 75, 80, 104, 124, 125, 131, 137, 138, 140, 142, 143, 148, 151, 153, 158, 159, 166, 171
community service, 171
competition, 126, 159
competitive grant program, 135
complexity, 19, 72, 84, 140, 152, 167
composition, 11
compounds, 159
computer, 90
conference, 11, 33, 37, 116
congress, 26
Congress, viii, 1, 2, 4, 11, 15, 16, 17, 21, 24, 26, 27, 28, 37, 46, 47, 60, 77, 81, 88, 110, 111, 112, 114, 116, 135
connectivity, 70, 132
consensus, 78, 131
conservation, 18, 55, 67, 105, 106, 109, 115, 117, 130, 154, 155, 157, 164, 165, 174
conservation programs, 18
constituents, 138
construction, 75, 127, 131
consumers, 143
consumption, 30, 32, 37, 53, 87, 101, 125, 137, 138, 139, 143
contaminant, 152, 158
contaminated water, 67
contamination, 3, 11, 24, 32, 37, 48, 104, 105, 113, 117, 125, 137, 142, 153, 155, 175
content analysis, 90
contingency, 151
control measures, 39, 104, 150, 154
controversial, 20
cooling, 33, 79
cooperation, 134
cooperative agreements, 45, 123, 173
Cooperative State Research, Education, and Extension Service, 111
coordination, 6, 14, 15, 16, 17, 83, 108, 115, 134, 147, 149, 165, 166, 168, 169, 171, 172
copper, 124
cost, viii, 2, 3, 4, 7, 20, 22, 23, 24, 39, 41, 67, 78, 134, 144, 150

criticism, 22
crop(s), 67
crop rotations, 67
crowding out, 112
cultural heritage, 124
curricula, 71, 108, 109, 166, 169, 170, 172
cycling, 137

D

damages, 4
database, 46, 54, 90, 116
decision makers, 102, 109, 145, 165, 166, 169
decision-making process, 13
deforestation, 161
degradation, 26, 37, 126, 143, 159
Department of Agriculture, 40, 57, 85, 106, 110, 111, 113, 115, 158
Department of Commerce, 40, 110, 113, 115, 120
Department of Defense, 40, 110, 113, 115
Department of Energy, 113
Department of Health and Human Services, 40, 110, 115, 120
Department of Homeland Security, 40, 110, 115, 120
Department of the Interior, 33, 40, 41, 48, 57, 59, 85, 110, 112, 113, 115, 120
Department of Transportation, 40, 60, 110, 115, 120
Departments of Agriculture, 33, 84, 85, 115
deposition, 20, 76, 79, 105, 113, 114, 117, 125, 131, 137, 152, 157, 175
deposits, 124
depth, 72, 110, 153
destruction, 126, 159
detection, 54, 71, 102, 103, 129, 145, 147
disadvantaged youth, 142
disbursement, 111
discharges, 137
diseases, 106, 160
dissolved oxygen, 174
distribution, 123, 135
diversity, 124, 161, 163
DNA, 150
draft, 22, 26, 56, 58, 59, 72, 73, 80, 84, 86, 87, 89
drinking water, 3, 11, 19, 28, 32, 33, 37, 113, 117, 124, 131, 132, 138, 143, 152, 159, 175
drought, 75, 78

E

ecological indicators, 22
ecological processes, 59, 126
ecology, 164
economic damage, viii, 29, 33, 78, 125, 151
ecosystem restoration, 4, 8, 15, 22, 23, 24, 27, 48
education, 38, 39, 41, 71, 101, 107, 108, 109, 117, 149, 166, 168, 169, 170, 171, 172, 175
educational institutions, 71, 109, 170
educators, 71, 171
election, 18, 133
e-mail, 85, 89, 90
endangered, 3, 70, 75, 106, 107, 111, 160, 161, 162, 164
endangered species, 3, 70, 75, 106, 107, 160, 161, 162, 164
energy, 79
enforcement, 66, 115, 122
engineering, 112, 114
environment(s), 23, 37, 43, 109, 113, 114, 124, 125, 138, 143, 159, 165, 167, 170
environmental change, 31, 59, 84
environmental conditions, 31, 39, 52, 60, 72, 78, 104, 140, 153, 159
environmental issues, 140
environmental protection, 147, 150
Environmental Protection Agency, viii, 2, 9, 10, 25, 26, 27, 31, 33, 40, 85, 113, 120, 121, 173
environmental quality, 23, 167
environmental stress, 152
environmental sustainability, 134
erosion, 55, 67, 104, 105, 106, 117, 154, 157, 158, 159, 174

Everglades, 22, 23, 57, 113
evidence, 35, 91
evolution, 24
Executive Order, 33, 110, 127, 132
exercise, 147
expenditures, 16, 125
expertise, 34, 88, 89, 115
exploitation, 132
exposure, 101, 139
extinction, 164

F

families, ix, 121
Farm Bill, 106, 157
fauna, ix, 121
feces, 113
federal advisory, 111
federal agency, viii, 2, 5, 13, 17, 33, 37, 133
federal authorities, viii, 2, 5, 46
federal funds, 34
federal government, vii, 1, 3, 4
Federal Highway Administration, 10, 40, 110, 115
field trials, 65
financial, 23, 34, 80, 85, 123
financial reports, 34, 85
financial resources, 123
Fish and Wildlife Service, 9, 10, 15, 33, 40, 41, 46, 70, 110, 111, 115, 130, 150, 158, 164
fisheries, 23, 36, 69
fishing, 32, 112, 124, 147, 149, 151
flame, 138
flame retardants, 138
flavor, 143
flexibility, 47
flooding, 78
flora, ix, 121
fluctuations, 126, 159
food, 109, 113, 117, 126, 144, 151, 159, 175
food chain, 113, 117, 151, 175
food web, 126, 144, 159
force, 81, 84
fouling, 149
freshwater, viii, 3, 29, 32, 66, 124, 149
fund transfers, 45
funds, viii, 2, 5, 7, 8, 11, 13, 18, 26, 28, 34, 35, 37, 38, 40, 41, 42, 44, 63, 67, 68, 70, 72, 74, 85, 88, 89, 91, 109, 111, 123, 129, 135, 136, 173

G

GAO, viii, 4, 13, 14, 17, 25, 26, 29, 30, 31, 36, 39, 40, 41, 42, 44, 45, 57, 77, 89, 109, 110, 111, 112, 113, 114, 115
genetics, 164
goods and services, 111
governance, 4, 11, 13, 14, 15, 16
governments, vii, 1, 3, 16, 26, 34, 71, 86, 91, 108, 125, 135, 138, 140, 143, 158, 168
GPRA, 31, 46, 51, 73, 111, 173
grant programs, viii, 2, 9
grants, 7, 8, 11, 15, 18, 26, 45, 51, 62, 74, 80, 89, 111, 123, 135, 136, 138, 173
grass(s), 67, 112
grazing, 67
greed, 54
green alga, 174
greenhouse, 79
greenhouse gas, 79
greenhouse gas emissions, 79
groundwater, 67, 132, 137
grouping, 91
growth, 76, 105, 155, 166, 174
guidance, 26, 30, 48, 56, 57, 58, 87, 122
guidelines, 11

H

harbors, 117, 158, 174
harmful effects, vii, 1, 4
harvesting, 161
hazardous waste, 101, 138, 140, 170
Health and Human Services, 33, 40, 84, 110, 115, 120, 132
health problems, 76

high school, 71
history, ix, 121, 124, 168
homes, 77
host, 64, 168
House, 25, 26, 32, 33, 115, 120
House of Representatives, 32
Housing and Urban Development, 33, 110, 115, 120, 132
human, vii, 1, 26, 71, 76, 104, 113, 117, 138, 151, 153, 159, 167, 175
human health, 76, 104, 113, 117, 153, 159, 175
hunting, 23, 112, 149
hybrid, 15

I

identification, 36, 54, 87, 89
imagery, 117, 174
impairments, 72, 101, 102, 104, 105, 107, 116, 138, 139, 143, 153, 155, 160, 165
improvements, 23, 28, 52, 73, 117, 132, 173, 174
individuals, 34, 37, 85, 88, 91, 116
industry(s), 32, 33, 125, 143, 149, 151, 163
infrastructure, 11, 14, 17, 25, 31, 74, 76, 77, 78, 79, 80, 83, 84, 107, 122, 128, 131, 140, 158, 168, 173
initiation, 161
insecticide, 76
institutions, 34, 71, 86, 91, 109, 113, 117, 170, 172, 175
insulators, 79
integration, 11, 80
integrity, vii, 2, 5, 25, 33, 36, 80, 101, 125, 128, 133, 139, 143, 165, 167, 171
interest groups, 88, 116
invasions, 39, 122, 129, 131
investment(s), 37, 38, 60, 71, 108, 166, 168
islands, 124
issues, vii, 1, 4, 5, 6, 11, 20, 24, 25, 26, 34, 51, 73, 79, 81, 108, 128, 129, 132, 134, 135, 140, 158, 160, 166, 167, 168, 172, 173

J

job training, 142
justification, 80, 85, 133

K

kill, 65

L

lakes, 28, 32, 33, 36, 70, 72, 73, 76, 107, 113, 114, 123, 124, 125, 127, 130, 132, 137, 144, 145, 152, 167, 168
land acquisition, 127
landscape(s), 106, 124, 160
laws, vii, 1, 4, 5, 7, 11, 15, 74
lead, viii, 2, 5, 13, 16, 24, 30, 31, 33, 49, 52, 53, 78, 82, 83, 87, 106, 132, 160
leadership, 13, 123
learning, 39, 88, 108, 122, 130, 166, 169, 171
legislation, viii, 2, 11
light, 34, 51, 61
liquidate, 111
loans, 80
local government, 34, 86, 91, 125, 134, 135, 138, 140, 143

M

majority, 62, 77, 138
marsh, 68
materials, 71, 125, 159
matter, 34, 35, 47, 52, 53, 54, 72, 73, 76, 86, 88, 89, 91, 116
measurements, 21
media, 108, 169
medical, 143
mercury, 30, 32, 53, 63, 112, 131, 125, 137, 138, 143
messages, 89
methodology, 35, 85, 103, 147

migration, 150, 151
mission, 112
Mississippi River, 145, 151
misuse, 132
models, 59, 151
modifications, 15, 27, 50
mortality, 164
Moses, 27
museums, 115, 117, 166, 171, 175

N

National Academy of Sciences, 112
National Park Service, 10, 40, 50, 72, 110, 115
National Research Council, 48, 78, 112
native species, 20, 33, 39, 73, 78, 124, 144, 147, 165
natural resource management, 56, 115, 126, 147
natural resources, 48, 113
Natural Resources Conservation Service, 10, 18, 40, 67, 110, 115, 158
natural shorescapes, ix, 121
Netherlands, 113
next generation, ix, 121, 166, 173
NGOs, 134
nitrogen, 20
NOAA, 9
North America, viii, 29, 32, 112, 124, 130
nuisance, 20, 103, 104, 105, 144, 153, 156, 172
nutrient(s), 20, 55, 67, 75, 104, 105, 106, 110, 117, 126, 152, 154, 155, 157, 158, 159, 161, 163, 174

O

Obama, vii, 1, 5, 33, 121, 127, 172
Obama Administration, vii, 1, 5
Office of Management and Budget, 11, 16, 26, 28, 46, 135
Office of the Inspector General, 34, 110
officials, ix, 30, 33, 34, 41, 46, 47, 48, 49, 50, 51, 52, 53, 54, 60, 61, 62, 63, 66, 70, 71, 72, 73, 74, 79, 80, 82, 86, 87, 88, 90, 110, 112, 115
oil, ix, 121, 164
oil spill, 164
operations, 158
opportunities, 14, 23, 69, 70, 104, 108, 124, 126, 135, 136, 142, 153, 159, 166, 168, 169
organize, 15
outreach, 39, 50, 71, 107, 108, 122, 130, 149, 168, 169, 170
overlap, 14, 15, 135
oversight, viii, 2, 7, 18, 19, 51, 136, 165
ownership, 172
oxygen, 110, 159, 174

P

Pacific, 57, 113
participants, 122
pathogens, 104, 152, 153, 159
pathways, 144
PCBs, 53, 87, 102, 117, 137, 138, 140, 141, 142, 143, 174
peer review, 133
performance indicator, 48
permit, 75
pesticide, 106, 157
pharmaceuticals, 138, 143
phosphorous, 52, 105, 117
phosphorus, 20, 36, 110, 113, 117, 154, 158, 174, 175
phytoplankton, 143
plankton, 151
plants, 33, 53, 69, 71, 104, 109, 112, 124, 153
pneumonia, 113
policy, 5, 20, 22, 25, 39, 56, 59, 102, 112, 113, 122, 123, 124, 129, 131, 145, 172
pollinators, 159
pollutants, 4, 19, 76, 79, 104, 125, 127, 137, 143, 152, 154

pollution, vii, viii, 2, 4, 5, 11, 29, 38, 39, 41, 48, 52, 55, 79, 100, 104, 105, 121, 124, 125, 129, 137, 138, 140, 141, 152, 154, 155, 158, 172
polychlorinated biphenyl (PCBs), 30, 31, 32, 87, 137
polycyclic aromatic hydrocarbon, 142
population, 76, 109, 113, 124, 151, 161, 164
population growth, 76
potential benefits, 4, 56
power plants, 33, 53
precipitation, 52, 76, 117, 174
predation, 164
President, 7, 26, 33, 37, 111, 112, 116, 121, 123, 127, 128, 131, 133, 172
President Obama, 33, 121
prevention, 39, 60, 71, 102, 121, 126, 141, 143
principles, 56, 82, 123, 133
probability, 109, 167, 169, 171
professional development, 166
profit, 164
progress reports, 51, 59, 60, 61, 88, 112, 113, 133
project, 6, 7, 15, 16, 18, 19, 21, 23, 27, 35, 40, 41, 49, 50, 51, 63, 64, 66, 68, 70, 71, 73, 74, 75, 80, 82, 84, 86, 88, 91, 102, 107, 112, 114, 123, 133, 134, 135, 136, 141, 142, 168
protection, vii, 2, 5, 6, 11, 16, 30, 38, 39, 41, 43, 46, 47, 48, 50, 62, 66, 71, 75, 80, 101, 108, 109, 111, 122, 126, 127, 131, 132, 135, 143, 147, 150, 152, 159, 162, 164, 167, 169, 170
public concern(s), 170
public health, 32, 48, 67, 83, 105, 155
public interest, 121
public policy, 112
public support, 6

Q

quality assurance, 73, 74
quality control, 171
quality standards, 104, 153
questionnaire, 89

R

rainfall, 77
raw materials, 125
recommendations, 5, 11, 15, 16, 17, 30, 35, 84, 85, 88, 135
recovery, 79, 83, 106, 107, 126, 140, 160, 161, 162, 163, 164, 175
recovery plan, 106, 160, 175
recreation, ix, 3, 103, 104, 115, 121, 124, 125, 148, 149, 152, 153
recreational, viii, 23, 29, 32, 36, 69, 70, 102, 124, 125, 142, 144, 145, 149, 152, 164
regional economies, 151
Registry, 10, 40, 110, 115
regulatory agencies, 167
rehabilitation, 126, 145, 161
reliability, 51
remedial actions, 105, 155
remediation, 11, 18, 24, 37, 105, 126, 138, 142, 154
remote sensing, 108, 109, 167, 169
reproduction, 143
requirements, 18, 51, 73, 74, 114
resilience, 21, 78
resource management, 14, 56, 115, 126, 147
resources, ix, 6, 14, 33, 36, 48, 62, 79, 80, 105, 106, 109, 113, 121, 123, 127, 131, 132, 134, 140, 151, 154, 160, 166, 167, 169, 171, 172
response, vii, 1, 3, 21, 31, 33, 35, 37, 56, 60, 71, 73, 89, 90, 102, 103, 113, 114, 116, 129, 131, 142, 145, 147, 148, 149, 150, 151
restoration programs, 4, 7, 8
restored ecosystem, viii, 2, 20, 22, 126
restrictions, 30, 36, 53, 87, 101, 139, 143
risk(s), 78, 102, 104, 117, 131, 137, 143, 145, 149, 150, 151, 153, 158, 174
rotations, 67
runoff, 39, 67, 76, 105, 117, 122, 137, 152, 154, 172, 174

S

safety, 151
sample survey, 89
sampling error, 89
school, 71, 117, 166, 175
science, 6, 16, 39, 56, 89, 108, 109, 112, 122, 123, 126, 133, 134, 165, 169, 171
Science Advisory Board, 37, 47, 52, 53, 54, 56, 58, 86, 111, 116
scope, viii, 2, 11, 14, 17, 18, 26, 30, 31, 35, 40, 73, 75, 76, 84, 85, 122
sea level, 78
sediment(s), 3, 4, 11, 18, 23, 24, 26, 34, 37, 41, 63, 67, 70, 78, 101, 102, 104, 105, 112, 113, 116, 117, 125, 129, 137, 138, 140, 141, 142, 152, 154, 157, 158, 159, 163, 174, 175
sedimentation, 117, 159, 174
Senate, 8, 26
sensing, 108, 109, 167, 169
sensitivity, 129
services, 111, 114, 163
severe stress, ix, 121
sewage, 78
shade, 79
shellfish, 76
shoreline, 3, 35, 68, 110, 112, 117, 152, 153, 164, 167, 174
shores, 124
showing, ix, 121
signs, ix, 121, 155
silver, 109
society, 56, 58, 114
soil erosion, 67, 104, 154, 158
solution, 23
specialists, 89
spending, 15, 28, 125
state laws, 74
storms, 78
stormwater, 31, 74-79, 83, 84, 152
strategic planning, 108, 134, 169
stress, ix, 121, 125, 127, 155, 172
stressors, 152
structure, viii, 2, 11, 13, 14, 85, 142
subsistence, 142
substrate, 167
Superfund, 116, 122, 138, 142, 174
surfactants, 138
surging, ix, 121
surveillance, 54, 103, 129, 143, 147, 151
survival, 164
survival rate, 164
sustainability, 113, 134
sustainable development, 4, 128, 173
symptoms, 25
synchronize, 140

T

target, 48, 55, 62, 67, 71, 103, 129, 130, 136, 147, 148, 158, 161, 167
Task Force agencies, ix, 29, 30, 31, 33, 34, 35, 37, 38, 40, 44, 45, 46, 50, 53, 55, 56, 58, 59, 60, 61, 72, 73, 75, 81, 82, 84, 85, 86, 87, 88, 110, 111, 115, 148
teacher training, 166, 172
teachers, 71, 166, 171
technical comments, 85
techniques, 36, 67
technology(s), 49, 60, 87, 103, 107, 108, 112, 129, 145, 147, 149, 167, 168, 169
telephone, 89, 90
temperature, 52, 117, 174
Tennessee Valley Authority, 113
testing, 48, 49, 56, 60, 66, 105, 149, 155
threats, 3, 81, 125, 127, 129, 138, 143, 150, 159, 164, 175
time frame, 50, 52, 82, 128
tissue, 113, 117, 175
toxic substances, vii, 1, 2, 4, 5, 26, 32, 38, 41, 53, 87, 100, 101, 117, 125, 137, 139, 141, 152, 174
toxicity, 72
tracks, 30, 52, 53
trade, 144
training, 71, 142, 166
transaction costs, 6, 134
transparency, 170
transport, 77, 159

transportation, 124
Treasury, 111
treatment, 23, 28, 60, 66, 71, 77, 131, 132, 145, 149, 151
tumbling tributaries, ix, 121
tumors, 62, 143

U

U.S. Army Corps of Engineers, 3, 9, 10, 40, 41, 57, 110, 113, 115, 117, 150, 158
U.S. Department of Agriculture, 57, 106, 113, 158
U.S. Department of Commerce, 113, 120
U.S. Department of the Interior, 57, 112, 113, 120
U.S. Geological Survey, 10, 40, 49, 57, 63, 64, 110, 115, 150, 158
U.S. Treasury, 111
unforeseen circumstances, 24
United, v, vii, 1, 3, 24, 25, 26, 29, 32, 34, 36, 43, 55, 56, 71, 72, 76, 81, 108, 109, 111, 113, 114, 115, 124, 125
United Nations, 72, 113
United States, v, vii, 1, 3, 24, 25, 26, 29, 32, 34, 36, 43, 55, 56, 71, 76, 81, 108, 109, 111, 114, 115, 124, 125
universe, 55, 89, 112
universities, 5, 46, 115, 117, 164, 175
updating, 11, 55, 58
urban, 3, 39, 104, 105, 107, 122, 137, 140, 142, 152, 154, 159, 160, 165, 172
urban areas, 104, 154
USDA, 157

V

vector, 144
vegetation, 67
vessels, 66
viruses, 104, 153
vision, 19, 20, 22
vulnerability, 80

W

wages, 32
Washington, 25, 26, 28, 32, 57, 58, 109, 110, 111, 112, 114, 115, 116
waste, 67, 101, 102, 138, 140, 141, 143, 170
wastewater, 28, 31, 71, 74, 76, 77, 78, 79, 83, 84, 131
water quality, vii, ix, 1, 3, 4, 14, 17, 67, 70, 76, 78, 80, 104, 113, 117, 121, 126, 152, 153, 161, 163, 165, 175
water quality standards, 104, 153
water resources, 109, 167, 169, 171
watershed, 2, 56, 58, 62, 105, 106, 111, 117, 145, 151, 152, 154, 155, 157, 158, 159, 170, 174
waterways, 33, 102, 123, 144, 145, 149, 151, 152
wealth, 124
web, 35, 88, 89, 90, 91, 159
wetlands, 7, 19, 21, 38, 39, 41, 67, 70, 72, 75, 107, 112, 113, 117, 122, 124, 140, 159, 160, 162, 163, 167, 175
White House, 25, 33, 115, 120
wilderness, 124
wildlife, vii, 1, 2, 4, 5, 6, 30, 36, 38, 39, 41, 46, 47, 50, 69, 75, 101, 106, 111, 124, 125, 126, 137, 139, 143, 159, 160, 162, 163, 165, 172
wind speed, 52
Wisconsin, 3, 15, 25, 32, 35, 86, 88, 142
worldwide, 114

Y

yield, 6

Z

zooplankton, 143